DESIGN AND IMPLEMENTATION OF LARGE-RANGE COMPLIANT MICROPOSITIONING SYSTEMS

DESIGN AND IMPLEMENTATION OF LARGE-RANGE COMPLIANT MICROPOSITIONING SYSTEMS

Qingsong Xu

University of Macau, China

Library of Congress Cataloging-in-Publication Data

Names: Xu, Qingsong, 1978- author.
Title: Design and implementation of large-range compliant micropositioning
    systems / Qingsong Xu.
Description: Singapore : John Wiley & Sons Inc., [2016] | Includes
    bibliographical references and index.
Identifiers: LCCN 2016015350| ISBN 9781119131434 (cloth) | ISBN 9781119131465
    (Adobe PDF) | ISBN 9781119131458 (epub) | ISBN 9781119131441 (oBook)
Subjects: LCSH: Micropositioners.
Classification: LCC TJ223.P67 X825 2016 | DDC 629.8/95–dc23 LC record available at
https://lccn.loc.gov/2016015350

A catalogue record for this book is available from the British Library.

Typeset in 10/12pt TimesLTStd by SPi Global, Chennai, India
Printed and bound in Singapore by Markono Print Media Pte Ltd

10 9 8 7 6 5 4 3 2 1

To Professor Wei Zhao and my family, for their constant encouragement

Contents

Preface   **xiii**

Acknowledgments   **xvii**

1 **Introduction**   **1**
1.1   Micropositioning Techniques   1
1.2   Compliant Guiding Mechanisms   2
       1.2.1   Basic Flexure Hinges   2
       1.2.2   Translational Flexure Hinges   3
       1.2.3   Translational Positioning Mechanisms   4
       1.2.4   Rotational Positioning Mechanisms   8
       1.2.5   Multi-Stroke Positioning Mechanisms   10
1.3   Actuation and Sensing   11
1.4   Control Issues   12
1.5   Book Outline   14
       References   14

**Part I   LARGE-RANGE TRANSLATIONAL MICROPOSITIONING
           SYSTEMS**

2 **Uniaxial Flexure Stage**   **21**
2.1   Concept of MCPF   21
       2.1.1   Limitation of Conventional Flexures   21
       2.1.2   Proposal of MCPF   23
2.2   Design of a Large-Range Flexure Stage   25
       2.2.1   Mechanism Design   25
       2.2.2   Analytical Modeling   26
       2.2.3   Architecture Optimization   29
       2.2.4   Structure Improvement   31
2.3   Prototype Development and Performance Testings   33
       2.3.1   Statics Performance Testing   34
       2.3.2   Dynamics Performance Testing   35

2.4	Sliding Mode Controller Design	35
	2.4.1 Dynamics Modeling	35
	2.4.2 DSMC Design	36
2.5	Experimental Studies	38
	2.5.1 Plant Model Identification	38
	2.5.2 Controller Setup	39
	2.5.3 Set-Point Positioning Results	39
	2.5.4 Sinusoidal Positioning Results	41
2.6	Conclusion	42
	References	44

3	**XY Flexure Stage**	**45**
3.1	Introduction	45
3.2	XY Stage Design	46
	3.2.1 Decoupled XY Stage Design with MCPF	46
	3.2.2 Buckling/Bending Effect Consideration	49
	3.2.3 Actuation Issues	51
3.3	Model Verification and Prototype Development	52
	3.3.1 Performance Assessment with FEA Simulation	52
	3.3.2 Prototype Fabrication	54
	3.3.3 Open-Loop Experimental Results	54
3.4	EMPC Control Scheme Design	55
	3.4.1 Problem Formulation	56
	3.4.2 EMPC Scheme Design	57
	3.4.3 State Observer Design	60
	3.4.4 Tracking Error Analysis	61
3.5	Simulation and Experimental Studies	61
	3.5.1 Plant Model Identification	61
	3.5.2 Controller Parameter Design	64
	3.5.3 Simulation Studies and Discussion	64
	3.5.4 Experimental Results and Discussion	66
3.6	Conclusion	67
	References	69

4	**Two-Layer XY Flexure Stage**	**70**
4.1	Introduction	70
4.2	Mechanism Design	71
	4.2.1 Design of a Two-Layer XY Stage with MCPF	71
	4.2.2 Structure Improvement of the XY Stage	72
4.3	Parametric Design	73
	4.3.1 Motion Range Design	73
	4.3.2 Stiffness and Actuation Force Design	74
	4.3.3 Critical Load of Buckling	75
	4.3.4 Resonant Frequency	75
	4.3.5 Out-of-Plane Payload Capability	76
	4.3.6 Influences of Manufacturing Tolerance	77

4.4	Experimental Studies and Results	79
	4.4.1 Prototype Development	80
	4.4.2 Statics Performance Testing	80
	4.4.3 Dynamics Performance Testing	81
	4.4.4 Positioning Performance Testing	83
	4.4.5 Contouring Performance Testing	84
	4.4.6 Control Bandwidth Testing	86
	4.4.7 Discussion and Future Work	88
4.5	Conclusion	89
	References	89

Part II   MULTI-STROKE TRANSLATIONAL MICROPOSITIONING SYSTEMS

5	**Dual-Stroke Uniaxial Flexure Stage**	**93**
5.1	Introduction	93
5.2	Mechanism Design and Analysis	94
	5.2.1 Mechanism Design to Minimize Interference Behavior	94
	5.2.2 Mechanism Design to Achieve Large Stroke	99
	5.2.3 FEA Simulation and Design Improvement	101
5.3	Prototype Development and Open-Loop Testing	104
	5.3.1 Experimental Setup	106
	5.3.2 Statics Performance Testing	106
	5.3.3 Dynamics Performance Testing	107
5.4	Controller Design and Experimental Studies	109
	5.4.1 Controller Design	109
	5.4.2 Experimental Studies	110
5.5	Conclusion	111
	References	113

6	**Dual-Stroke, Dual-Resolution Uniaxial Flexure Stage**	**114**
6.1	Introduction	114
6.2	Conceptual Design	115
	6.2.1 Design of a Compliant Stage with Dual Ranges	115
	6.2.2 Design of a Compliant Stage with Dual Resolutions	116
6.3	Mechanism Design	117
	6.3.1 Stiffness Calculation	118
	6.3.2 Motion Range Design	119
	6.3.3 Motor Stroke and Driving Force Requirement	120
	6.3.4 Sensor Deployment	121
6.4	Performance Evaluation	123
	6.4.1 Analytical Model Results	123
	6.4.2 FEA Simulation Results	124
6.5	Prototype Development and Experimental Studies	125
	6.5.1 Prototype Development	126
	6.5.2 Statics Performance Testing	127

	6.5.3	Dynamics Performance Testing	129
	6.5.4	Further Discussion	131
6.6	Conclusion		133
	References		133

7	**Multi-Stroke, Multi-Resolution XY Flexure Stage**		**135**
7.1	Introduction		135
7.2	Conceptual Design		136
	7.2.1	Design of Flexure Stage with Multiple Strokes	136
	7.2.2	Design of Flexure Stage with Multiple Resolutions	138
7.3	Flexure-Based Compliant Mechanism Design		139
	7.3.1	Compliant Element Selection	139
	7.3.2	Design of a Two-Axis Stage	140
7.4	Parametric Design		141
	7.4.1	Design of Motion Strokes	141
	7.4.2	Design of Coarse/Fine Sensor Resolution Ratio	144
	7.4.3	Actuation Issue Consideration	145
7.5	Stage Performance Assessment		146
	7.5.1	Analytical Model Evaluation Results	146
	7.5.2	FEA Simulation Results	146
7.6	Prototype Development and Experimental Studies		149
	7.6.1	Prototype Development	149
	7.6.2	Statics Performance Testing	150
	7.6.3	Dynamics Performance Testing	154
	7.6.4	Circular Contouring Testing	156
	7.6.5	Discussion	156
7.7	Conclusion		159
	References		159

Part III   LARGE-RANGE ROTATIONAL MICROPOSITIONING SYSTEMS

8	**Rotational Stage with Linear Drive**		**163**
8.1	Introduction		163
8.2	Design of MCRF		164
	8.2.1	Limitation of Conventional Radial Flexures	164
	8.2.2	Proposal of MCRF	165
	8.2.3	Analytical Models	166
8.3	Design of a Rotary Stage with MCRF		169
	8.3.1	Consideration of Actuation Issues	170
	8.3.2	Consideration of Sensing Issues	172
8.4	Performance Evaluation with FEA Simulation		172
	8.4.1	Analytical Model Results	172
	8.4.2	FEA Simulation Results	173
	8.4.3	Structure Improvement	175

8.5	Prototype Development and Experimental Studies	176
	8.5.1 Prototype Development	176
	8.5.2 Open-Loop Performance Testing	177
	8.5.3 Controller Design and Closed-Loop Performance Testing	178
	8.5.4 Further Discussion	181
8.6	Conclusion	183
	References	184
9	**Rotational Stage with Rotary Drive**	**185**
9.1	Introduction	185
9.2	New Design of MCRF	186
	9.2.1 MCRF Design	186
	9.2.2 Analytical Model Not Considering Deformation	187
	9.2.3 Analytical Model Considering Deformation	189
9.3	Design of the Rotary Stage	192
	9.3.1 Actuator Selection	194
	9.3.2 Sensor Design	194
9.4	Performance Evaluation with FEA Simulation	196
	9.4.1 Analytical Model Results	197
	9.4.2 FEA Simulation Results	197
9.5	Prototype Fabrication and Experimental Testing	201
	9.5.1 Prototype Development	201
	9.5.2 Statics Performance Testing	202
	9.5.3 Dynamics Performance Testing	206
	9.5.4 Discussion	206
9.6	Conclusion	207
	References	208

Part IV   APPLICATIONS TO COMPLIANT GRIPPER DESIGN

10	**Large-Range Rotary Gripper**	**213**
10.1	Introduction	213
	10.1.1 Structure Design and Driving Method	213
	10.1.2 Sensing Requirements	214
10.2	Mechanism Design and Analysis	216
	10.2.1 Actuation Issues	216
	10.2.2 Position and Force Sensing Issues	218
10.3	Performance Evaluation with FEA Simulation	222
	10.3.1 Analytical Model Results	222
	10.3.2 FEA Simulation Results	222
10.4	Prototype Development and Calibration	227
	10.4.1 Prototype Development	227
	10.4.2 Calibration of Position Sensor	228
	10.4.3 Calibration of Force Sensor	229

	10.4.4	Verification of Force Sensor	230
	10.4.5	Consistency Testing of the Sensors	231
10.5	Performance Testing Results		232
	10.5.1	Testing of Gripping Sensing Performance	232
	10.5.2	Testing of Horizontal Interaction Detection	235
	10.5.3	Testing of Vertical Interaction Detection	236
	10.5.4	Testing of Dynamics Performance	237
	10.5.5	Applications to Pick–Transport–Place in Assembly	238
	10.5.6	Further Discussion	239
10.6	Conclusion		242
	References		242

11	**MEMS Rotary Gripper**		**244**
11.1	Introduction		244
11.2	MEMS Gripper Design		245
	11.2.1	Actuator Design	246
	11.2.2	Sensor Design	249
11.3	Performance Evaluation with FEA Simulation		251
	11.3.1	Statics Analysis	252
	11.3.2	Dynamics Analysis	254
11.4	Gripper Fabrication		254
11.5	Experimental Results and Discussion		255
	11.5.1	Gripping Range Testing Results	255
	11.5.2	Gripping Force Testing Results	258
	11.5.3	Interaction Force Testing Results	260
	11.5.4	Demonstration of Micro-object Gripping	261
	11.5.5	Further Discussion	262
11.6	Conclusion		264
	References		266

| **Index** | | | **267** |

Preface

Micropositioning systems refer to positioning devices which are able to produce displacement down to sub-micrometer resolution and accuracy. Such devices are widely employed to realize a precise positioning of microrobotic end-effectors dedicated to precision manipulation and assembly applications. To cater for the precision requirement in relatively low-loading scenarios, flexure-based compliant mechanisms have been exploited extensively owing to their attractive merits in terms of no backlash, no friction, no wear, low cost, and vacuum compatibility. Unlike traditional mechanical joints, the repeatable output motion of a flexure mechanism is delivered via the elastic deformation of the material.

Typically, flexure mechanisms can deliver a translational displacement of less than 1 mm and a rotational displacement smaller than 1°. In modern precision engineering applications, there is a growing demand for micropositioning systems which are capable of providing large-range precision motion (e.g., over 10 mm translation and 10° rotation), yet possess a compact size at the same time. Such applications range from scanning probe microscopy to wafer alignment, lithography and fabrication, biological micromanipulation, etc. A precision positioning stage with a compact size allows the application inside a limited space. Additionally, a compact physical size enables cost reduction in terms of material and fabrication. For practical applications, once the kinematic scheme is determined, the structural parameters of the flexure mechanism need to be carefully designed to make sure that the material operates in the elastic domain without plastic deformation or fatigue failure.

Traditionally, the motion range is restricted by the mechanism design - due to the stress concentration and stress stiffening effects - and also constrained by the maximum allowable stress of the material. Intuitively, a larger motion range can be achieved by employing flexures with longer and more slender hinges. However, the length of the flexure hinge is usually constrained by the compactness requirement and the minimum width is restricted by the tolerance of the manufacturing process in practice. Hence, it is challenging to design a flexure micropositioning stage with a large stroke and compact size simultaneously. To this end, this book is concentrated on the design and development of flexure-based compact micropositioning systems with large motion ranges. Some innovative mechanism designs are presented for large-range translational and rotational positioning. Analytical modeling and finite-element analysis are carried out to evaluate the performance of the mechanisms. Prototypes have been developed for experimental investigations.

To implement a complete micropositioning system, suitable actuation and sensing schemes are selected. Once a micropositioning device is constructed by incorporating the flexure micropositioning stage, sensors, and actuators properly, its accuracy is dependent on a suitable control strategy. Usually, a micropositoining system is termed a nanopositioning system if it can provide the displacement resolution down to sub-nanometer or nanometer level. As typical control schemes, the proportional-integral-derivative (PID), sliding mode control (SMC), and model predictive control (MPC) algorithms are realized as examples to achieve a precise positioning of the micropositoining systems in this book.

The book also involves the design of large-range compliant grippers, which combine the large-range translational and rotational stages together. The realization of the gripper down to microelectromechanical systems (MEMS) scale is also demonstrated. Detailed examples of their analyses and implementations are provided. A comprehensive treatment of the subject matter is provided in a manner amenable to readers ranging from researchers to engineers, by providing detailed simulation and experimental verifications of the developed devices.

The book begins with an introduction to micropositioning techniques and provides a brief survey of development and applications in Chapter 1. According to the different implementations of micropositioning systems, the remaining ten chapters of the book are divided into four parts.

Part I consists of Chapters 2, 3, and 4, which address the design and implementation of large-range translational micropositioning systems. Specifically, Chapter 2 presents the design of a uniaxial translational positioning device by introducing the idea of multi-stage compound parallelogram flexure (MCPF). A voice coil motor (VCM) and a laser displacement sensor are adopted for the actuation and sensing of the developed stage, respectively. Control experiments are demonstrated to verify the stage performance. Chapters 3 and 4 devise large-range, parallel-kinematic, decoupled XY micropositioning systems, which can provide two-dimensional decoupled translations over 10 mm in each working axis. Several variations of the decoupled XY flexure stage are designed. While Chapter 3 proposes a monolithic structure design, Chapter 4 reports on a two-layer compact design of the parallel-kinematic XY flexure mechanism.

Part II is composed of Chapters 5, 6, and 7, which present multi-stroke translational micropositioning systems. Chapter 5 describes the design and implementation of a flexure-based dual-stage micropositioning system. A VCM and a fine piezoelectric stack actuator (PSA) are adopted to provide the long stroke and quick response, respectively. A decoupling design is proposed to minimize the interference behavior between the coarse and fine stages by taking into account the actuation schemes as well as guiding mechanism implementations. Chapters 6 and 7 propose the conceptual design of multi-stroke, multi-resolution uniaxial and two-dimensional micropositioning stages, respectively, which are driven by a single actuator for each working axis. The stages are devised based on a fully compliant variable-stiffness mechanism, which exhibits unequal stiffnesses in different strokes. Resistive strain sensors are employed to offer variable displacement resolutions in different strokes.

Part III includes Chapters 8 and 9, which deal with the design and implementation of large-range rotational micropositioning systems. Based on the idea of multi-stage compound radial flexure (MCRF), two kinds of rotary compliant stages are devised to achieve both a large rotational range over 10° and a compact size. Chapter 8 presents a rotational micropositioning device which is driven by a linear VCM and sensed by a laser displacement sensor, whereas Chapter 9 reports a rotational micropositioning system which is actuated by a rotary VCM and measured by a strain-gauge sensor. Analytical models are derived to facilitate the parametric designs, which are validated by conducting finite-element analysis (FEA) simulations. Experimental results reveal a large rotational output motion of the developed rotational devices with a low level of center shift.

As a typical application of the presented translational and rotational micropositioning stages, Part IV proposes the design and development of innovative large-range compliant grippers. Chapter 10 devises a compliant gripper with integrated position and force sensors dedicated to automated robotic microhandling tasks. The gripper is capable of detecting grasping force and environmental interaction forces in the horizontal and vertical axes. Moreover, a variable-stiffness compliant mechanism is designed to provide the force sensing with dual sensitivities and dual measuring ranges. Chapter 11 reports a realization of the compliant gripper in MEMS scale. The gripper is driven by an electrostatic actuator and measured by a capacitive sensor. The integrated gripper possesses a compact size, less than 4 mm × 6 mm, and is fabricated using the silicon-on-insulator (SOI) microfabrication technique. The performance of the gripper is demonstrated via experimental studies.

This book provides state-of-the-art coverage of the methodology of compliant mechanisms for achieving large-range translational and rotational positioning in the context of mechanism design, analytical modeling, drive and sensing, motion control, and experimental testing. Detailed examples of their implementations are provided. Readers can expect to learn how to design and develop new flexure-based compliant micropositioning systems to realize large-range translational or rotational motion dedicated to precision engineering applications.

Acknowledgments

The author would like to acknowledge the University of Macau (under Grants SRG006-FST11-XQS, MYRG083(Y1-L2)-FST12-XQS, and MYRG078(Y1-L2)-FST13-XQS) and the Science and Technology Development Fund (FDCT) of Macao (under Grants 024/2011/A, 070/2012/A3, and 052/2014/A1) for co-funding the projects. The author is also grateful for the help provided by Ms. Stephanie Loh and Ms. Maggie Zhang from John Wiley.

1

Introduction

Abstract: This chapter presents a brief introduction of micropositioning systems and their concerned design and control problems. The compliant translational and rotational guiding mechanisms are described, the related actuation and sensing issues are raised, and the motion control problem is summarized. An outline of the remaining chapters of the book is provided.

Keywords: Micropositioning, Compliant mechanisms, Flexure hinges, Translational guiding, Rotational guiding, Actuators, Sensors, Control.

1.1 Micropositioning Techniques

Micropositioning systems refer to precision positioning devices which are capable of delivering displacement down to sub-micrometer resolution and accuracy. Micropositioning devices have been widely applied in the domain of precision manipulation and manufacturing, such as scanning probe microscopy, lithography manufacturing, and wafer alignment. To cater for the precision demands in relatively low-loading applications, flexure-based compliant mechanisms have been widely employed. Unlike traditional mechanical joints, the repeatable output motion of a flexible element is generated by the elastic deformation of the material. As a consequence, compliant mechanisms enable some attractive advantages – including no backlash, no friction, no wear, low cost, vacuum compatibility, etc. [1, 2].

According to the motion property, micropositioning can be classified into two general categories in terms of translational and rotational micropositioning. The combination of these two types of motion forms a hybrid micropositioning. Typical flexure mechanisms can deliver a translational displacement of less than 1 mm and a rotational displacement smaller than 1° within the yield strength of the materials. In modern precision engineering applications, there is a growing demand for micropositioning systems which are capable of producing large-range (e.g., over 10 mm or 10°) precision motion, yet have a compact size at the same time. Such applications involve large-range scanning probe microscopy [3], lithography and fabrication [4], biological micromanipulation [5], etc. For instance, in automated zebrafish embryo manipulation, a precise positioning stage with a long stroke is needed to execute accurate operation [6].

Design and Implementation of Large-Range Compliant Micropositioning Systems, First Edition. Qingsong Xu.
© 2016 John Wiley & Sons Singapore Pte Ltd. Published 2016 by John Wiley & Sons Singapore Pte Ltd.

In addition, a precision positioning stage with compact size allows the application inside a constrained space. For example, a compact positioning device is required to provide ultrahigh-precision positioning of the specimens and tools inside the chamber of scanning electron microscopes for automated probing and micromanipulation [7]. Moreover, a compact physical size enables cost reduction in terms of material and fabrication. Hence, this book is concentrated on the design and implementation of compact micropositioning stages with large motion ranges.

1.2   Compliant Guiding Mechanisms

Concerning the motion guiding mechanism of the positioning stage, although aerostatic bearings [8] and maglev bearings [9] are usually adopted, flexure bearings are more attractive in the recent development of micropositioning systems, due to the aforementioned merits of compliant mechanisms [10]. Compared with other mechanisms, compliant flexures can generate a smooth motion by making use of the elastic deformation of the material. Nevertheless, their motion range is constricted by the yield strength of the material, which poses a great challenge to achieving a long stroke. From this point of view, once the kinematic scheme is determined, the structural parameters of the flexure mechanism call for a careful design to make sure that the material operates in the elastic domain without plastic deformation and fatigue failure.

Given the requirements on the motion or force property, a compliant guiding mechanism can be designed by resorting to different approaches, such as the rigid-body replacement method [11], building-block method [12], topology optimization method [13], topology synthesis method [14], etc. Without loss of generality, the element flexure hinges and the translational and rotational positioning mechanisms are introduced in the following sections.

1.2.1   Basic Flexure Hinges

A basic flexure hinge functions as a revolute joint. In the literature, various profiles of flexure hinges have been used to construct a flexure stage [15]. For example, the in-plane profiles of typical flexure hinges including right-circular, elliptic, right-angle, corner-filled, and leaf hinges are shown in Fig. 1.1. More types of flexure hinges are referred to in the books [2, 16].

Referring to Fig. 1.1, if one terminal A of the flexure hinge is fixed and the other terminal B has an applied force F_x along the x-axis or a moment M_z around the z-axis, an in-plane bending deformation of the hinge will be induced. Generally, these element flexure hinges are considered as revolute joints, which deliver a rotational motion of the terminal B with respect to the fixed terminal A around a rotation center. To generate a translational motion or a multi-axis rotational motion like a universal or spherical joint, multiple basic flexure hinges can be combined to construct a compound flexure hinge [17].

During the bending deformation of the element flexure hinge, the rotation center will be varied. The notch-type flexure hinge, especially the right-circular hinge, is able to deliver a rotation with smaller amount of center shift. However, this is achieved at the cost of a relatively small rotational motion range due to the stress concentration effect. In order to accomplish a large motion range, the leaf flexure hinge is usually employed due to the mitigation of the stress concentration effect. In addition, leaf flexures have been widely employed in micromechanism

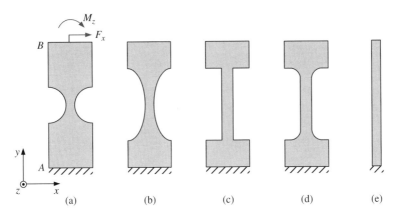

Figure 1.1   Profiles of typical flexure hinges: (a) right-circular hinge; (b) elliptic hinge; (c) right-angle hinge; (d) corner-filled hinge; (e) leaf hinge.

design in microelectromechanical systems (MEMS) devices [18]. The design methods of the beam-based leaf flexures are referred to in the book [1].

1.2.2   *Translational Flexure Hinges*

As a compound type of flexure, parallelogram flexure is a popular design to achieve translational motion. For example, the translational flexure hinges constructed by right-circular hinges are shown in Fig. 1.2. To generate a larger translational motion range, the translational flexure hinges can be designed using leaf hinges, as shown in Fig. 1.3.

As shown in Fig. 1.3(a), when the output stage of a parallelogram flexure translates a displacement d_x in the x-axis, it also undergoes a parasitic translation d_y in the y-axis. For some applications, the translation d_y can be employed to enhance the resolution of the displacement due to the displacement deamplification effect. Concerning a large-range positioning in the

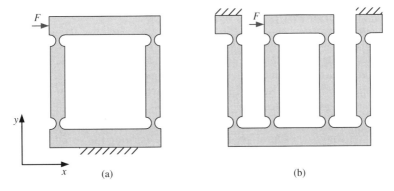

Figure 1.2   Translational flexure hinges constructed by right-circular hinges: (a) parallelogram flexure; (b) compound parallelogram flexure (CPF).

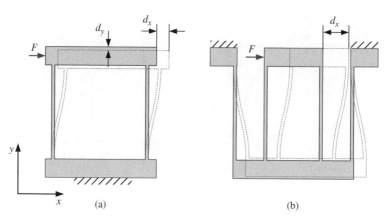

Figure 1.3 Translational flexure hinges constructed by leaf hinges: (a) parallelogram flexure; (b) compound parallelogram flexure (CPF).

specified direction, the parasitic translation d_y is unwanted. In order to obtain a larger straight motion while eliminating the parasitic translation, a compound parallelogram flexure (CPF), as shown in Fig. 1.3(b), can be employed.

Intuitively, a longer stroke can be realized by using a longer and more slender leaf flexure. However, in practice, the length of the flexure hinge is constrained by the requirement of compactness and the minimum width is restricted by the tolerance of the manufacturing process. It is challenging to design a flexure micropositioning stage with a large stroke and compact size simultaneously. To overcome the aforementioned problem, the concept of multi-stage compound parallelogram flexure (MCPF) [19], as shown in Fig. 1.4(a), is employed in this book.

Compared with conventional CPF, the motion range of a MCPF is enlarged N times without changing the length and width of the flexures, where N is the number of basic CPF modules. Note that CPF is a special case of MCPF with $N = 1$. To enhance the transverse stiffness in the y-axis direction, an improved MCPF is presented as shown in Fig. 1.4(b), which is constructed by connecting the two secondary stages together.

1.2.3 Translational Positioning Mechanisms

A translational positioning mechanism is usually required to provide the translational motion in the two-dimensional plane or three-dimensional space. To generate the translational positioning in more than one direction, a suitable mechanism design is necessary. As far as a kinematic scheme is concerned, the positioning stages, which are capable of multi-dimensional translations, can be classified into two categories in terms of serial and parallel kinematics. The majority of the commercially available stages employ a serial-kinematic scheme. For example, some micropositioning stages have been developed by stacking the second single-axis positioning stage on top of the first one or nesting the second stage inside the first one [20–22]. In this way, the entire second stage is carried by the first one, as illustrated in Fig. 1.5(a), where the X stage serves as the output platform of the XY stage. As an example, the computer-aided design (CAD) model of a serial-kinematic XY stage is shown in Fig. 1.6(a), where the parallelogram flexures are constructed using right-circular hinges.

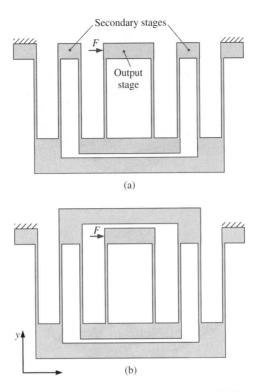

(a)

(b)

Figure 1.4 (a) A multi-stage compound parallelogram flexure (MCPF) with two modules; (b) an improved MCPF with enhanced transverse stiffness in the *y*-axis.

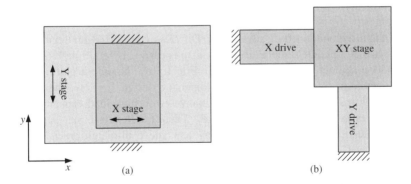

(a) (b)

Figure 1.5 Illustrations of (a) serial-kinematic XY stage and (b) parallel-kinematic XY stage.

Even though a compact structure may be achieved by using the serial-kinematic design [22], it is at the cost of high inertia, low resonant frequency, and large cumulative errors. A further disadvantage is that the dynamic characteristics in the different working axes are usually unequal for a serial-kinematic stage. On the contrary, a parallel-kinematic scheme [23, 24]

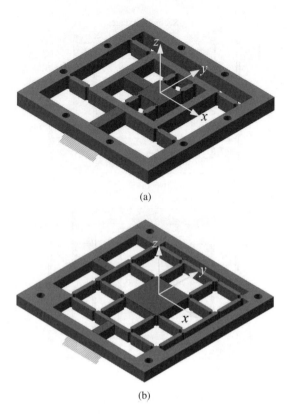

(a)

(b)

Figure 1.6   Examples of (a) serial-kinematic XY compliant stage and (b) parallel-kinematic XY compliant stage.

overcomes the aforementioned disadvantages. Different from serial-kinematic mechanisms, the end-effector of a parallel-kinematic mechanism is carried by multiple kinematic linkages in a closed-loop manner [25]. As illustrated in Fig. 1.5(b), the output platform is driven by X and Y drives in parallel. Unlike the serial-kinematic design, it allows the achievement of low inertia, high resonant frequency, no cumulative error, high load capacity, and identical dynamic features in the different working axes. Thus, the flexure-based parallel-kinematic compliant mechanisms pave a promising way to achieve ultrahigh-precision positioning. For instance, the CAD model of a parallel-kinematic XY stage is shown in Fig. 1.6(b). Although the right-circular hinges are adopted as examples to construct the parallelogram flexures, any other types of hinges (e.g., leaf flexures) can also be employed to design the XY stage.

To facilitate the control design for the micropositioning systems, the micropositioning stages are desirable to provide a decoupled output motion. Output decoupling means that the output motion in one working axis does not induce motion in the other axes of the stage. Additionally, input decoupling indicates that the actuation provided by one motor does not cause a force or load on the other motors of the stage. The purpose of input decoupling is to isolate and protect the actuators for a micropositioning system. A total decoupling stage possesses the properties of both output decoupling and input decoupling. The XY stage shown in Fig. 1.6(b) is desired to possess total decoupling characteristics. However, such a standard flexure-based XY micropositioning stage is restricted to deliver a small translational range, partially because of the stress stiffening effect.

Stress stiffening is a geometrical nonlinearity. It is most pronounced in structures which are thin in one or more dimensions. Given a structure based on flexure hinges as shown in Fig. 1.1, the stress stiffening indicates that the lateral stiffness in the x-axis of the structure can be significantly increased (or decreased) by the tensile (or compressive) axial stress in the y-axis of the structure. Generally, this phenomenon should be avoided because it increases the actuation force requirement and reduces the stroke of the motor, and causes nonlinearities in actuation. This book presents the design of large-range micropostioning systems with the stress stiffening effect mitigated greatly.

Recently, some compliant XY stages have been proposed to deliver a large motion range over 1 mm [26–29]. However, the developed stages have a relatively large dimension. As a result, the stages possess a small area ratio, which is defined as the ratio between the area of the planar workspace and the area of the planar dimension of the XY stage. To achieve a large motion range while keeping a compact structure, the MCPFs are proposed to devise new compliant parallel-kinematic XY stages in this book.

In addition, as a spatial mechanism, a traditional XYZ micropositioning stage is shown in Fig. 1.7. This XYZ stage is called a three-prismatic-universal-universal (3-PUU) parallel mechanism [30]. The cube-like output platform is supported by three identical limbs, which are arranged orthogonally and connected in parallel. Each limb consists of a serial connection of one prismatic (P) hinge and two universal (U) hinges. Each universal hinge includes two orthogonally arranged notch-type revolute (R) hinges. The XYZ stage delivers a nearly decoupled output translation in the three-dimensional space. However, the motion range is limited due to the relatively small rotational angle of the notch-type flexure hinges.

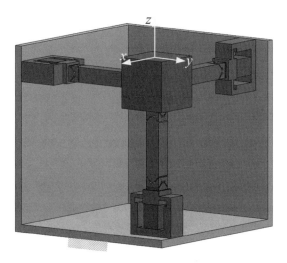

Figure 1.7   Example of a parallel-kinematic XYZ compliant stage.

1.2.4   Rotational Positioning Mechanisms

Translational micropositioning mechanisms have drawn the attention of numerous researchers [19] because they are relatively easy to implement. Nevertheless, for many scenarios such as semiconductor manufacturing, microalignment devices, and optics devices [31], a micropositioning system which is capable of precision rotary positioning is required. Unfortunately, only limited previous work can be found in this category.

In the literature, compliant stages providing combined translational and rotational motions have been reported [32–34]. For example, Fig. 1.8 shows a planar three-revolute-revolute-revolute (3-RRR) flexure parallel mechanism, which can provide two translational motions in the *xy*-plane and one rotational motion around the *z*-axis. Spatial compliant mechanisms have also been reported to deliver coupled translations and rotations in three-dimensional space [35]. Additionally, precision stages with spherical motions have been developed [36]. This book involves the design and implementation of rotational compliant micropositioning stages which are capable of pure rotary motion. Such rotary stages have found extensive applications in precision engineering. Several rotational flexure stages have been proposed in previous work [37–40]. However, the majority of existing stages are only able to deliver a small rotary angle less than 1°. In practice, a rotational stage with a larger angle is demanded in many situations. How to achieve a large rotary range by using flexure-based compliant mechanisms is a major challenge.

A conventional compliant rotational stage is illustrated in Fig. 1.9. The rotational stage is guided by radial flexures with fixed–guided constraint [39, 41, 42], as shown in Fig. 1.10(a). The circle-like output platform can rotate around the center of the stage. However, its rotational range is limited due to the mechanism overconstraint and stress stiffening effect [43]. To enlarge the rotary angle, several rotational bearings have been presented [44–46] and some rotational stages driven by smart material-based actuators (e.g., piezoelectric actuator and shape memory alloy) have been devised [47]. For instance, a butterfly-shaped flexure pivot is reported [44], which exhibits the characteristics of small parasitic center translation and monolithic structure. More recently, a large-displacement compliant

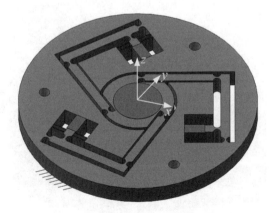

Figure 1.8   Example of a flexure-based compliant 3-RRR parallel stage.

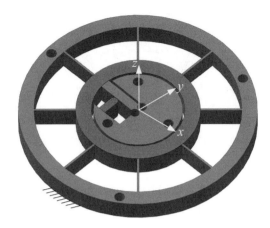

Figure 1.9   Example of a flexure-based compliant rotational stage.

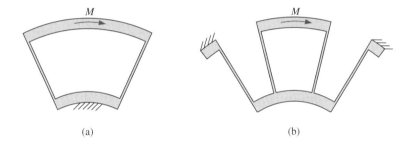

Figure 1.10   (a) A radial flexure; (b) a compound radial flexure (CRF).

rotational hinge called Flex-16 has been proposed [46], which is able to rotate 90° without failure.

Alternatively, the basic module of radial flexure can be employed to construct compound radial flexures (CRFs) [48], as illustrated in Fig. 1.10(b). To achieve a large rotational range, the CRFs should be designed with larger length, smaller thickness, and larger outer radius. However, these physical parameters are restricted by the compactness constraint, manufacturing tolerance, and requirement on the minimum stiffness in practice. Thus, it is difficult to achieve a large rotational range while maintaining a compact stage size by using CRFs.

To cope with the above problem, the concept of multi-stage compound radial flexure (MCRF) [49] is employed in this book to devise a rotational stage with enlarged rotary angle as well as compact physical dimension. One kind of MCRF with two modules is shown in Fig. 1.11(a). Furthermore, to enhance the transverse stiffness in the radial direction (toward the rotation center), an improved MCRF is given in Fig. 1.11(b), which is constructed by connecting the two secondary stages together.

A survey on the recent development of large-stroke compliant micropositioning stages has been reported in the literature [50].

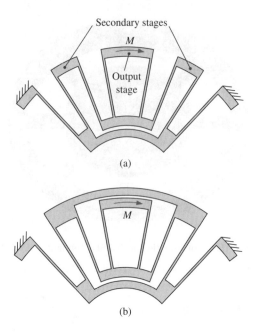

Figure 1.11   (a) A multi-stage compound radial flexure (MCRF) with two modules; (b) an improved MCRF with enhanced stiffness in the radial direction.

1.2.5   Multi-Stroke Positioning Mechanisms

Multi-stroke micropositioning is highly desirable in scenarios where the merits of high positioning accuracy, long stroke motion, and high servo bandwidth are simultaneously required. Usually, a dual-servo system (DSS) is employed to deliver a coarse motion and a fine motion together. In applications such as scanning probe microscopy (SPM), a precision nanopositioning stage is used to implement an accurate and rapid raster scanning operation to get surface profiles of the scanned specimen [51, 52]. Nevertheless, only a small portion of the specimen can be put on the scanning table. For instance, an atomic force microscope (AFM) usually delivers a scanning range less than 200 μm × 200 μm. In order to acquire the surface topography of a large specimen (e.g., 10 mm × 10 mm), a nanopositioning stage with both a large workspace and a high bandwidth is required to fully cover the whole specimen surface and to quickly acquire the surface profile of interested scanning areas, as illustrated in Fig. 1.12. Thus, the DSS opens the way to cater for these demands. DSSs also have promising applications in biological micromanipulation [53] and microgripping [54].

Most of the existing DSSs are applied in hard-disk drives [55]. Recently, DSSs have been extended to micro-/nanopositioning applications. For example, a dual-stage nanopositioning system is developed in [56], where a coarse permanent magnet stepper motor stage and a fine piezoelectric stack actuator stage are stacked together. A coarse/fine dual-stage system is reported in [57] by using a voice coil motor to drive a fine stage and a permanent magnet linear synchronous motor to drive a coarse stage. Additionally, the design of a linear dual-stage actuation system is presented [8] by employing a voice coil motor and a piezoelectric stack actuator as the coarse and fine drivers, respectively. In the foregoing work, aerostatic bearings

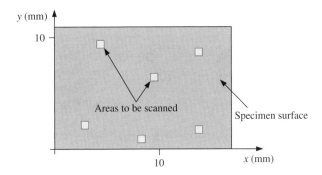

Figure 1.12   Illustration of required scan area over the surface of a large specimen.

and maglev bearings are most adopted to guide the output motion of the stage. By contrast, flexure bearings are preferred due to their merits of no backlash, no friction, vacuum compatibility, easy manufacturing, and so on. Thus, flexures are employed in the recent development of dual-stage micro-/nanopositioning systems [58–60].

The major issue of a dual-servo stage arises from the interference behavior, which is caused by the interaction between the coarse and fine stages. In this book, the idea of mechanical decoupling design is introduced to minimize the interference behavior. In addition, one actuator conventionally only delivers a single stroke along with a specific resolution. It is challenging to devise a single-drive stage with multiple strokes as well as multiple resolutions. In this book, different conceptual designs of multi-stroke flexure-based micropositioning systems will be presented.

1.3   Actuation and Sensing

The micropositioning stage is usually called a nanopositioning stage if it can deliver a motion resolution at (sub-)nanometer level. Because the compliant mechanism can deliver a smooth and repeatable motion, the generation of ultrahigh-level resolution for flexure-based micropositioning systems is dependent on the selection of actuators and sensors.

Piezoelectric stack actuators (PSAs) are widely applied in micro-/nanopositioning stages. However, PSAs typically deliver a short stroke up to 0.1% of their length [61]. Although various lever transmission mechanisms can be employed to amplify the output displacement [62, 63], it is practically difficult to realize a positioning over 10 mm. In order to achieve a large motion range, the ball-screw drives are mostly employed [64]. Such actuation produces a large motion range yet introduces nonlinearity in terms of the friction effect, which may deteriorate the positioning accuracy. In addition, the stick–slip actuators have been reported to provide unlimited strokes in theory [65]. However, the stick–slip motion is usually achieved by resorting to the friction force, which may induce mechanical wear and block the achievement of continuous and accurate motion. Hence, some friction-free drives have been employed, such as magnetic levitation motors [66], electromagnetic actuators [67], and voice coil motors (VCMs) [68, 49].

These types of motors are lubrication free and vacuum compatible, hence also fulfilling applications in ultraclean environments. Without loss of generality, the VCMs are employed

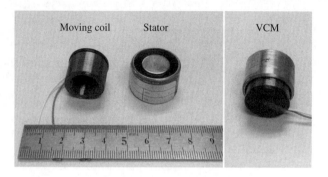

Figure 1.13   A linear voice coil motor (VCM).

to drive the large-range micropositioning stages. The VCM is a linear or rotary actuator, which consists of a permanent magnetic stator and a moving cylinder encompassed by a coil of copper wire. A photograph of a linear VCM (model: NCC04-10-005-1A, from H2W Techniques, Inc.) and its components is shown in Fig. 1.13, which can provide a linear stroke of 10.2 mm. The VCM works based on Lorentz force law. That is, a current-carrying wire in a magnetic field creates a force which is perpendicular to the directions of the current and the magnetic field. Recent applications of VCMs in high-precision positioning stages are reviewed in [69]. In this book, both linear and rotary VCMs are used to actuate the translational and rotational micropositioning systems.

Concerning the sensing issue, the output displacement of a micropositioning device can be measured by various displacement sensors based on different sensing principles [70]. In the literature, optical sensors, capacitive sensors, strain sensors, and inductance sensors have been employed in micropositioning systems. In this book, the laser displacement sensors, capacitive sensors, and strain-gauge sensors are selected as examples in the prototype implementations. To achieve nanometer-level resolution in a large positioning range for the large-range micropositioning systems, laser interferometers can be adopted.

In this book, the micropositioning systems are designed to provide the specified large-range translational or rotational motion in determinate directions. Depending on the materials, the micropositioning stages can be fabricated by resorting to wire-electrical discharge machining (EDM), laser cutting, and water-jet cutting processes. As an alternative type of compliant mechanism, soft mechanisms have recently been developed to deliver the large-range motion in an arbitrary direction [71]. Such a robotic mechanism is usually designed based on soft and deformable materials and driven by pneumatic, hydraulic, and smart material-based actuators, etc. Concerning manufacture, the soft mechanisms are usually fabricated by means of rapid prototyping (RP), 3-D printing, and so on.

1.4   Control Issues

Once the micropositioning stage is fabricated, the achievement of precision positioning is dependent on the control technique. Due to the low damping of the flexure-based system, a number of vibration modes exist and a high-order model is commonly identified for the system plant, which may result in a high-order controller as a consequence. From the implementation point of view, a linear model of lower order (e.g., second order) is more desirable for practical realization of the control algorithm. However, the adoption of a low-order model means that the residual modes are not considered. Challenges exist in control design because the neglect of residual modes may cause control spillover and observation spillover [72]. Spillover is undesirable as it may cause instability and performance degradation of the system [73]. In addition, to facilitate the control design, model nonlinearity including the hysteresis effect is usually considered as a lumped disturbance to the nominal plant model. The suppression of the disturbance requires a robust control approach.

To account for these problems, various control techniques have been developed to maintain the robustness of the system in the presence of model uncertainties and disturbances. The block diagram of a typical feedback control is shown in Fig. 1.14. Nowadays, the most popular control approach in various industrial domains is proportional-integral-derivative (PID) control [74]. One reason for this is that, as a model-free control scheme, PID is easy to maintain in practice. However, PID control exhibits weak robustness against system uncertainties and external disturbances. Hence, it cannot always produce satisfactory control results for some applications where the plants are accompanied by nonlinearities and disturbances. To achieve better performance for a micropositioning system, advanced control schemes have been employed, such as the gain scheduling control [75], loop shaping technique [76], H_∞ robust control [77], adaptive control [78], model-reference control [79], repetitive control [80], iterative learning control [81], fuzzy logic control [82], and neuro-fuzzy control [83], to name just a few.

Moreover, sliding mode control (SMC) has been demonstrated as a simple yet efficient non-linear robust control approach to tolerate model uncertainty and disturbance. The essence of SMC is to drive the system state trajectory onto a specified sliding mainfold and then keep it moving along the surface. Once the sliding surface is reached, the controlled system is robust against certain variation of the model and external disturbances. In recent years, a variety of SMC strategies have been developed for precision motion systems [84]. Furthermore, discrete-time sliding mode control (DSMC) is more feasible for implementation on sampled-data systems [85]. A DSMC scheme can be developed based on the system state or system output [86, 87]. In a typical micropositioning system, only the position information is supplied by the displacement sensors. In order to realize the DSMC scheme, a state observer is usually needed if the full state information is required [88]. Some recent DSMC strategies have been developed to eliminate the use of a state observer [89].

Additionally, as a robust control strategy, model predictive control (MPC) is well-known for its ability to solve the problems with constraints, time delays, and disturbances by offering an optimal control action [90]. The essence of the MPC scheme is to obtain the current

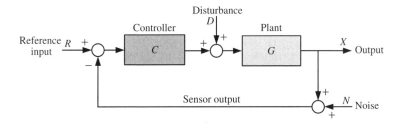

Figure 1.14 Block diagram of feedback control of a micropositioning system.

control action by optimizing the predictions of plant behavior over a finite sequence of future control inputs. At each sampling time instant, the MPC controller generates an optimal control sequence by solving an optimization problem and uses the first element of the sequence as the control action for the system plant. Recently, MPC schemes have been widely employed in the domain of precision motion control [87, 91].

Regarding micro-/nanopositioning systems driven by piezoelectric actuators, advanced control methodologies have been presented in a recent monograph [92]. In this book, typical control algorithms including PID, DSMC, and MPC are implemented to demonstrate the performance of the developed micropositioning systems.

1.5 Book Outline

According to the different implementations of the micropositioning systems, the remaining ten chapters of the book are divided into four parts. Part I presents the design and implementation of large-range translational micropositioning systems. Part II develops the design and development of multi-stroke translational micropositioning systems. Part III proposes the design and realization of large-range rotational micropositioning systems. As typical applications of the devised translational and rotational micropositioning systems, Part IV reports on the design and development of innovative large-range compliant grippers.

All the chapters involve both simulation and experimental verification studies. Moreover, further research directions following each chapter are proposed as potential future work. The presented design ideas can easily be extended to the design of various MEMS devices. Even though planar motion is mostly concerned in this book, the presented design methodologies can be expanded to the design of spatial compliant mechanisms to achieve three-dimensional large-range motion for pertinent applications.

References

[1] Howell, L.L. (2001) *Compliant Mechanisms*, Wiley, New York.

[2] Lobontiu, N. (2002) *Compliant Mechanisms: Design of Flexure Hinges*, CRC Press, Boca Raton, FL.

[3] Sinno, A., Ruaux, P., Chassagne, L., Topcu, S., Alayli, Y., Lerondel, G., *et al.* (2007) Enlarged atomic force microscopy scanning scope: Novel sample-holder device with millimeter range. *Rev. Sci. Instrum.*, **78** (9), 095 107.

[4] Lee, D.J., Kim, K., Lee, K.N., Choi, H.G., Park, N.C., Park, Y.P. *et al.* (2010) Robust design of a novel three-axis fine stage for precision positioning in lithography. *Proc. Inst. Mech. Eng. Part C-J. Mech. Eng. Sci.*, **224** (4), 877–888.

[5] Gauthier, M. and Piat, E. (2006) Control of a particular micro-macro positioning system applied to cell micromanipulation. *IEEE Trans. Automat. Sci. Eng.*, **3** (3), 264–271.

[6] Huang, H.B., Sun, D., Mills, J.K., and Cheng, S.H. (2009) Robotic cell injection system with position and force control: Toward automatic batch biomanipulation. *IEEE Trans. Robot.*, **25** (3), 727–737.

[7] Zhang, Y.L., Zhang, Y., Ru, C., Chen, B.K., and Sun, Y. (2013) A load-lock-compatible nanomanipulation system for scanning electron microscope. *IEEE/ASME Trans. Mechatron.*, **18** (1), 230–237.

[8] Dong, W., Tang, J., and ElDeeb, Y. (2009) Design of a linear-motion dual-stage actuation system for precision control. *Smart Mater. Struc.*, **18**, 095 035–1–095 035–11.

[9] Chen, M.Y., Wang, C.C., and Fu, L.C. (2001) Adaptive sliding mode controller design of a dual-axis maglev positioning system, in *Proc. American Control Conf.*, Arlington, VA, pp. 3731–3736.

[10] Hegde, S. and Ananthasuresh, G.K. (2012) A spring-mass-lever model, stiffness and inertia maps for single-input, single-output compliant mechanisms. *Mech. Mach. Theory*, **58**, 101–119.

[11] Howell, L.L. and Midha, A. (1995) Parametric deflection approximations for end-loaded, large-deflection beams in compliant mechanisms. *J. Mech. Des.*, **117** (1), 156–165.

[12] Bernardoni, P., Bidaud, P., Bidard, C., and Gosselin, F. (2004) A new compliant mechanism design methodology based on flexible building blocks, in *Proc. SPIE*, vol. 5383, San Diego, CA, vol. 5383, pp. 244–254.

[13] Wang, M.Y., Wang, X., and Guo, D. (2003) A level set method for structural topology optimization. *Comput. Meth. Appl. Mech. Eng.*, **192** (1), 227–246.

[14] Hopkins, J.B. and Culpepper, M.L. (2010) Synthesis of multi-degree of freedom, parallel flexure system concepts via freedom and constraint topology (FACT)-part I: Principles. *Precis. Eng.*, **34** (2), 259–270.

[15] Yong, Y.K., Lu, T.F., and Handley, D.C. (2008) Review of circular flexure hinge design equations and derivation of empirical formulations. *Precis. Eng.*, **32** (2), 63–70.

[16] Smith, S.T. (2000) *Flexures: Elements of Elastic Mechanisms*, Gordon and Breach, New York.

[17] Trease, B.P., Moon, Y.M., and Kota, S. (2004) Design of large-displacement compliant joints. *J. Mech. Des.*, **127** (4), 788–798.

[18] Krijnen, B., Swinkels, K.R., Brouwer, D.M., Abelmann, L., and Herder, J.L. (2015) A large-stroke 3DOF stage with integrated feedback in MEMS. *J. Microelectromech. Syst.*, **24** (6), 1720–1729.

[19] Xu, Q. (2012) New flexure parallel-kinematic micropositioning system with large workspace. *IEEE Trans. Robot.*, **28** (2), 478–491.

[20] Fung, R.F., Hsu, Y.L., and Huang, M.S. (2009) System identification of a dual-stage XY precision positioning table. *Precis. Eng.*, **33** (1), 71–80.

[21] Kenton, B.J. and Leang, K.K. (2012) Design and control of a three-axis serial-kinematic high-bandwidth nanopositioner. *IEEE/ASME Trans. Mechatron.*, **17** (2), 356–369.

[22] Wadikhaye, S., Yong, Y.K., and Moheimani, S.O.R. (2012) Design of a compact serial-kinematic scanner for high-speed atomic force microscopy: An analytical approach. *Micro Nano Lett.*, **7** (4), 309–313.

[23] Yong, Y.K. and Lu, T.F. (2009) Kinetostatic modeling of 3-RRR compliant micro-motion stages with flexure hinges. *Mech. Mach. Theory*, **44** (6), 1156–1175.

[24] Ramadan, A.A., Takubo, T., Mae, Y., Oohara, K., and Arai, T. (2009) Developmental process of a chopstick-like hybrid-structure two-fingered micromanipulator hand for 3-D manipulation of microscopic objects. *IEEE Trans. Ind. Electron.*, **56** (4), 1121–1135.

[25] Merlet, J.P. (2006) *Parallel Robots*, Springer, Berlin.

[26] Kang, D., Kim, K., Kim, D., Shim, J., Gweon, D.G., and Jeong, J. (2009) Optimal design of high precision XY-scanner with nanometer-level resolution and millimeter-level working range. *Mechatronics*, **19** (4), 562–570.

[27] Chen, M.Y., Huang, H.H., and Hung, S.K. (2010) A new design of a submicropositioner utilizing electromagnetic actuators and flexure mechanism. *IEEE Trans. Ind. Electron.*, **57** (1), 96–106.

[28] Awtar, S. and Parmar, G. (2010) Design of a large range XY nanopositioning system, in *Proc. ASME 2010 Int. Design Engineering Technical Conf.*, Montreal, Canada, pp. 387–399.

[29] Hao, G. and Kong, X. (2012) A novel large-range XY compliant parallel manipulator with enhanced out-of-plane stiffness. *J. Mech. Des.*, **134** (6), 061 009.

[30] Xu, Q. and Li, Y. (2006) Stiffness modeling for an orthogonal 3-PUU compliant parallel micromanipulator, in *Proc. IEEE Int. Conf. on Mechatronics and Automation*, Luoyang, Henan, China, pp. 124–129.

[31] Chong, J., He, S., and Ben Mrad, R. (2012) Development of a vector display system based on a surface-micromachined micromirror. *IEEE Trans. Ind. Electron.*, **59** (12), 4863–4870.

[32] Wang, H. and Zhang, X. (2008) Input coupling analysis and optimal design of a 3-DOF compliant micro-positioning stage. *Mech. Mach. Theory*, **43** (4), 400–410.

[33] Hwang, D., Byun, J., Jeong, J., and Lee, M.G. (2011) Robust design and performance verification of an in-plane XYθ micropositioning stage. *IEEE Trans. Nanotechnol.*, **10** (6), 1412–1423.

[34] Qin, Y., Shirinzadeh, B., Zhang, D., and Tian, Y. (2013) Design and kinematics modeling of a novel 3-DOF monolithic manipulator featuring improved Scott–Russell mechanisms. *J. Mech. Des.*, **135** (10), 101 004.

[35] Yang, G., Teo, T.J., Chen, I.M., and Lin, W. (2011) Analysis and design of a 3-DOF flexure-based zero-torsion parallel manipulator for nano-alignment applications, in *Proc. IEEE Int. Conf. on Robotics and Automation*, Shanghai, China, pp. 2751–2756.

[36] Tan, K.K., Huang, S., Liang, W., Mamun, A.A., Koh, E.K., and Zhou, H. (2012) Development of a spherical air bearing positioning system. *IEEE Trans. Ind. Electron.*, **59** (9), 3501–3509.

[37] Cannon, J.R. and Howell, L.L. (2005) A compliant contact-aided revolute joint. *Mech. Mach. Theory*, **40** (11), 1273–1293.

[38] Wang, Y.C. and Chang, S.H. (2006) Design and performance of a piezoelectric actuated precise rotary positioner. *Rev. Sci. Instrum.*, **77** (10), 105 101.

[39] Kim, K., Ahn, D., and Gweon, D. (2012) Optimal design of a 1-rotational DOF flexure joint for a 3-DOF H-type stage. *Mechatronics*, **22** (1), 24–32.

[40] Teo, T.J., Yang, G., and Chen, I.M. (2014) A large deflection and high payload flexure-based parallel manipulator for UV nanoimprint lithography: Part I. Modeling and analyses. *Precis. Eng.*, **38** (4), 861–871.

[41] Luo, H.P., Zhang, B., and Zhou, Z.X. (2008) A rotary flexural bearing for micromanufacturing. *CIRP Ann.-Manuf. Techn.*, **57**, 179–182.

[42] Bi, S., Zhao, S., and Zhu, X. (2010) Dimensionless design graphs for three types of annulus-shaped flexure hinges. *Precis. Eng.*, **34** (3), 659–666.

[43] Choi, Y.J., Sreenivasan, S.V., and Choi, B.J. (2008) Kinematic design of large displacement precision XY positioning stage by using cross strip flexure joints and over-constrained mechanism. *Mech. Mach. Theory*, **43** (6), 724–737.

[44] Henein, S., Spanoudakis, P., Droz, S., Myklebust, L.I., and Onillon, E. (2003) Flexure pivot for aerospace mechanisms, in *Proc. 10th European Space Mechanisms and Tribology Symp.*, San Sebastian, Spain, pp. 1–4.

[45] Pei, X., Yu, J., Zong, G., Bi, S., and Su, H. (2009) The modeling of cartwheel flexural hinges. *Mech. Mach. Theory*, **44** (10), 1900–1909.

[46] Fowler, R.M., Maselli, A., Pluimers, P., Magleby, S.P., and Howell, L.L. (2014) Flex-16: A large-displacement monolithic compliant rotational hinge. *Mech. Mach. Theory*, **82**, 203–217.

[47] Li, J., Zhao, H., Qu, H., Cui, T., Fu, L., Huang, H., *et al.* (2013) A piezoelectric-driven rotary actuator by means of inchworm motion. *Sens. Actuator A-Phys.*, **194**, 269–276.

[48] Valois, M. (2012), Rotary flexure bearing. US 2012/0034027 A1.

[49] Xu, Q. (2013) Design and implementation of a novel rotary micropositioning system driven by linear voice coil motor. *Rev. Sci. Instrum.*, **84** (5), 055 001.

[50] Wan, S. and Xu, Q. (2014) A survey on recent development of large-stroke compliant micropositioning stage. *Int. J. Robot. Automat. Technol.*, **1** (1), 19–35.

[51] Mokaberi, B. and Requicha, A.A.G. (2008) Compensation of scanner creep and hysteresis for AFM nanomanipulation. *IEEE Trans. Automat. Sci. Eng.*, **5** (2), 197–206.

[52] Leang, K.K., Zou, Q., and Devasia, S. (2009) Feedforward control of piezoactuators in atomic force microscope systems: Inversion-based compensation for dynamics and hysteresis. *IEEE Control Syst. Mag.*, **29** (1), 70–82.

[53] Gauthier, M. and Piat, E. (2006) Control of a particular micro-macro positioning system applied to cell micromanipulation. *IEEE Trans. Automat. Sci. Eng.*, **3** (3), 264–271.

[54] Rakotondrabe, M. and Ivan, I.A. (2011) Development and force/position control of a new hybrid thermo-piezoelectric microgripper dedicated to micromanipulation tasks. *IEEE Trans. Automat. Sci. Eng.*, **8** (4), 824–834.

[55] Al Mamun, A., Mareels, I., Lee, T.H., and Tay, A. (2003) Dual-stage actuator control in hard disk drive – a review, in *Proc. 29th Annual IEEE Industrial Electronics Society Conf.*, pp. 2132–2137.

[56] Michellod, Y., Mullhaupt, P., and Gillet, D. (2006) Strategy for the control of a dual-stage nano-positioning system with a single metrology, in *Proc. IEEE Conf. on Robotics, Automation and Mechatronics*, pp. 1–8.

[57] Song, Y., Wang, J., Yang, K., Yin, W., and Zhu, Y. (2010) A dual-stage control system for high-speed, ultra-precise linear motion. *Int. J. Adv. Manuf. Technol.*, **48**, 633–643.

[58] Dong, W., Sun, L.N., and Du, Z.J. (2007) Design of a precision compliant parallel positioner driven by dual piezoelectric actuators. *Sens. Actuator A-Phys.*, **135** (1), 250–256.

[59] Xu, Q. (2012) Design and development of a flexure-based dual-stage nanopositioning system with minimum interference behavior. *IEEE Trans. Automat. Sci. Eng.*, **9** (3), 554–563.

[60] Tuma, T., Haeberle, W., Rothuizen, H., Lygeros, J., Pantazi, A., and Sebastian, A. (2014) Dual-stage nanopositioning for high-speed scanning probe microscopy. *IEEE/ASME Trans. Mechatron.*, **19** (3), 1035–1045.

[61] Yao, Q., Dong, J., and Ferreira, P.M. (2007) Design, analysis, fabrication and testing of a parallel-kinematic micropositioning XY stage. *Int. J. Mach. Tools Manuf.*, **47** (6), 946–961.

[62] Kim, J.J., Choi, Y.M., Ahn, D., Hwang, B., Gweon, D.G., and Jeong, J. (2012) A millimeter-range flexure-based nano-positioning stage using a self-guided displacement amplification mechanism. *Mech. Mach. Theory*, **50**, 109–120.

[63] Xu, Q. and Li, Y. (2011) Analytical modeling, optimization and testing of a compound bridge-type compliant displacement amplifier. *Mech. Mach. Theory*, **46** (2), 183–200.

[64] Otsuka, J., Ichikawa, S., Masuda, T., and Suzuki, K. (2005) Development of a small ultraprecision positioning device with 5 nm resolution. *Meas. Sci. Technol.*, **16** (11), 2186.

[65] Breguet, J.M. and Clavel, R. (1998) Stick and slip actuators: Design, control, performances and applications, in *Proc. Int. Symp. on Micromechatronics and Human Science*, pp. 89–95.

[66] Verma, S., Kim, W.J., and Gu, J. (2004) Six-axis nanopositioning device with precision magnetic levitation technology. *IEEE/ASME Trans. Mechatron.*, **9** (2), 384–391.

[67] Kim, D.H., Lee, M.G., Kim, B., and Sun, Y. (2005) A superelastic alloy microgripper with embedded electro-magnetic actuators and piezoelectric force sensors: A numerical and experimental study. *Smart Mater. Struct.*, **14** (6), 1265–1272.

[68] Rakuff, S. and Cuttinob, J.F. (2009) Design and testing of a long-range, precision fast tool servo system for diamond turning. *Precis. Eng.*, **33** (1), 18–25.

[69] Shan, G., Li, Y., Zhang, L., Wang, Z., Zhang, Y., and Qian, J. (2015) Application of voice coil motors in high-precision positioning stages with large travel ranges. *Rev. Sci. Instrum.*, **86** (10), 101 501.

[70] Ouyang, P.R., Tjiptoprodjo, R.C., Zhang, W.J., and Yang, G.S. (2008) Micro-motion devices technology: The state of arts review. *Int. J. Adv. Manuf. Technol.*, **38** (5&6), 463–478.

[71] Rus, D. and Tolley, M.T. (2015) Design, fabrication and control of soft robots. *Nature*, **521** (7553), 467–475.

[72] Pai, M.C. and Sinha, A. (2000) Sliding mode control of vibration in a flexible structure via estimated states and H_∞/μ techniques, in *Proc. American Control Conf.*, Baltimore, MD, pp. 1118–1123.

[73] Wang, D.A. and Huang, Y.M. (2003) Application of discrete-time variable structure control in the vibration reduction of a flexible structure. *J. Sound Vibr.*, **261** (3), 483–501.

[74] Jaensch, M. and Lamperth, M.U. (2007) Investigations into the stability of a PID-controlled micropositioning and vibration attenuation system. *Smart Mater. Struc.*, **16** (4), 1066–1075.

[75] Lee, S.Q., Kim, Y., and Gweon, D.G. (2000) Continuous gain scheduling control for a micro-positioning system: Simple, robust and no overshoot response. *Control Eng. Pract.*, **8** (2), 133–138.

[76] Youm, W., Jung, J., Lee, S., and Park, K. (2008) Control of voice coil motor nanoscanners for an atomic force microscopy system using a loop shaping technique. *Rev. Sci. Instrum.*, **79** (1), 013 707.

[77] Dong, J., Salapaka, S.M., and Ferreira, P.M. (2008) Robust control of a parallel-kinematic nanopositioner. *J. Dyn. Sys., Meas., Control*, **130** (4), 041 007–1–041 007–15.

[78] Tan, X. and Baras, J.S. (2005) Adaptive identification and control of hysteresis in smart materials. *IEEE Trans. Automatic Control*, **50** (6), 827–839.

[79] Xu, Q. and Jia, M. (2014) Model reference adaptive control with perturbation estimation for a micropositioning system. *IEEE Trans. Contr. Syst. Technol.*, **22** (1), 352–359.

[80] Li, Y. and Xu, Q. (2012) Design and robust repetitive control of a new parallel-kinematic XY piezostage for micro/nanomanipulation. *IEEE/ASME Trans. Mechatron.*, **17** (6), 1120–1132.

[81] Wu, Y. and Zou, Q. (2007) Iterative control approach to compensate for both the hysteresis and the dynamics effects of piezo actuators. *IEEE Trans. Contr. Syst. Technol.*, **15** (5), 936–944.

[82] Chi, Z., Jia, M., and Xu, Q. (2014) Fuzzy PID feedback control of piezoelectric actuator with feedforward compensation. *Math. Probl. Eng.*, **2014**. Article ID 107184, doi: 10.1155/2014/107184.

[83] Lin, L.C., Sheu, J.W., and Tsay, J.H. (2001) Modeling and hierarchical neuro-fuzzy control for flexure-based micropositioning systems. *J. Intell. Robot. Syst.*, **32** (4), 411–435.

[84] Yu, X. and Kaynak, O. (2009) Sliding mode control with soft computing: A survey. *IEEE Trans. Ind. Electron.*, **56** (9), 3275–3285.

[85] Gao, W., Wang, Y., and Homaifa, A. (1995) Discrete-time variable structure control systems. *IEEE Trans. Ind. Electron.*, **42** (2), 117–122.

[86] Xu, J.X. and Abidi, K. (2008) Discrete-time output integral sliding-mode control for a piezomotor-driven linear motion stage. *IEEE Trans. Ind. Electron.*, **55** (11), 3917–3926.

[87] Xu, Q. and Li, Y. (2012) Model predictive discrete-time sliding mode control of a nanopositioning piezostage without modeling hysteresis. *IEEE Trans. Contr. Syst. Technol.*, **20** (4), 983–994.

[88] Ghafarirad, H., Rezaei, S.M., Abdullah, A., Zareinejad, M., and Saadat, M. (2011) Observer-based sliding mode control with adaptive perturbation estimation for micropositioning actuators. *Precis. Eng.*, **35** (2), 271–281.

[89] Xu, Q. (2014) Digital sliding-mode control of piezoelectric micropositioning system based on input–output model. *IEEE Trans. Ind. Electron.*, **61** (10), 5517–5526.

[90] Morari, M. and Lee, J.H. (1999) Model predictive control: Past, present and future. *Comput. Chem. Eng.*, **23** (4 & 5), 667–682.

[91] Xu, Q. (2015) Digital sliding mode prediction control of piezoelectric micro/nanopositioning system. *IEEE Trans. Contr. Syst. Technol.*, **23** (1), 297–304.

[92] Xu, Q. and Tan, K.K. (2015) *Advanced Control of Piezoelectric Micro-/Nano-Positioning Systems*, Springer, Berlin.

Part One

Large-Range Translational Micropositioning Systems

Part One

Large-Range Translational Micropositioning Systems

2

Uniaxial Flexure Stage

Abstract: This chapter presents the design, modeling, analysis, testing, and control of a compact long-stroke uniaxial flexure micropositioning stage. Parametric design is performed to achieve the highest natural frequency under certain constraints. A voice coil motor (VCM) and a laser displacement sensor are adopted for actuation and sensing of the fabricated stage, respectively. The conducted finite-element analysis (FEA) and experimental testing confirm a motion range up to 11 mm with a sub-micrometer accuracy. In addition, a discrete-time sliding mode control (DSMC) is implemented to enable a rapid and precise positioning in the presence of model uncertainties and disturbances. Experimental results show that the DSMC is superior to PID control in terms of transient response speed and steady-state accuracy.

Keywords: Uniaxial micropositioning, Flexure mechanisms, Translational motion, Finite-element analysis, Voice coil motor, Discrete-time sliding mode control, Motion control.

2.1 Concept of MCPF

In this section, the idea of multi-stage compound parallelogram flexure (MCPF) is introduced to achieve a large pure translational motion in one direction of motion.

2.1.1 Limitation of Conventional Flexures

In order to achieve a pure translational motion, compound parallelogram flexures (CPFs), as shown in Fig. 2.1(a) and (b), have been widely used [1, 2]. By examining the two types of CPFs, which have different connecting and fixing schemes, it is observed that both of them emerge from a common module as shown in Fig. 2.1(c).

When a force F_x is applied on the primary stage of the CPF, the deformed shape of the first type of CPF is depicted in Fig. 2.2(a). Each of the four flexures undergoes a combined force

Design and Implementation of Large-Range Compliant Micropositioning Systems, First Edition. Qingsong Xu.
© 2016 John Wiley & Sons Singapore Pte Ltd. Published 2016 by John Wiley & Sons Singapore Pte Ltd.

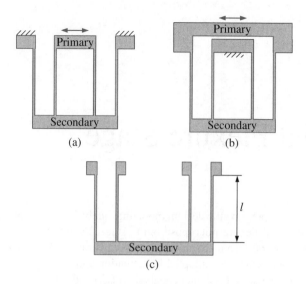

Figure 2.1   (a), (b) Ordinary compound parallelogram flexures; (c) the basic module.

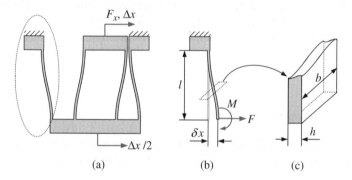

Figure 2.2   (a) Deformation of a compound parallelogram flexure; (b) bending deformation of one flexure; (c) dimensions of the cross-section.

F and moment M as shown in Fig. 2.2(b). Taking into account the boundary conditions of the rotation angle and translational motion, the following relationships can be derived:

$$0 = \frac{Fl^2}{2EI} - \frac{Ml}{EI} \tag{2.1}$$

$$\delta x = \frac{Fl^3}{3EI} - \frac{Ml^2}{2EI} \tag{2.2}$$

where δx is the translated displacement of one flexure, E is the Young's modulus of the material, and I is the area moment of inertia of the cross-section about the neutral axis:

$$I = \frac{bh^3}{12} \tag{2.3}$$

where h is the in-plane width and b is the out-of-plane thickness of the leaf flexure.

The force and displacement can be solved from Eqs. (2.1) and (2.2) as follows:

$$F = \frac{2M}{l} \tag{2.4}$$

$$\delta x = \frac{Fl^3}{12EI}. \tag{2.5}$$

Because the four flexures possess identical length (l), the relationship $\delta x = \Delta x/2$ holds, where Δx denotes the one-sided translation of the CPF. The stiffness of the CPF can be computed as

$$K^{CPF} = \frac{F_x}{\Delta x} = \frac{2F}{2\delta x} = \frac{Ebh^3}{l^3}. \tag{2.6}$$

When the flexures experience the maximum moment M_{max}, the maximum stress σ_{max}, which is dependent on the yield strength σ_y of the material, emerges on the outermost edge of some cross-section. It can be calculated via

$$\sigma_{max} = \frac{M_{max}h/2}{I} \tag{2.7}$$

which gives the following moment value in view of Eq. (2.3):

$$M_{max} = \frac{\sigma_{max}bh^2}{6}. \tag{2.8}$$

Taking into account Eqs. (2.4), (2.6), and (2.8), it can be derived that

$$\Delta x_{max}^{CPF} = \frac{F_{max}}{K^{CPF}} = \frac{\sigma_{max}l^2}{3Eh} \tag{2.9}$$

which indicates that the maximum one-sided translation Δx of the CPF is governed by the length l and in-plane width h of the leaf flexures, as well as the ratio σ_{max}/E of the material. To obtain a larger Δx, the material with a higher ratio of σ_{max}/E can be selected and the flexures are expected to have a longer length and smaller width.

The commonly used materials with relatively high ratio between the yield strength and Young's modulus are listed in the book [3]. For a selected material, the physical parameter l is restricted by the compactness requirement of the device, and h is limited by the manufacturing process and the minimum-stiffness requirement of the device. In the following discussions, the concept of MCPF is proposed to achieve a large translation while keeping the main parameters l and h unchanged.

2.1.2 Proposal of MCPF

The ordinary CPFs are constructed using a single module as shown in Fig. 2.1(c). To realize a larger motion range, the idea of MCPF with N modules is employed [4], as shown in Fig. 2.3(a). It is observed that the primary stage (i.e. the output stage) is connected to the base through N modules with multiple secondary stages. Additionally, to facilitate the design process, all of the leaf flexures are supposed to have identical length l.

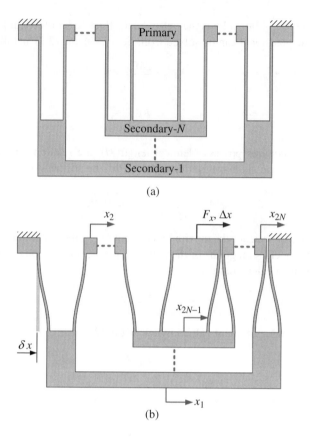

Figure 2.3   (a) A MCPF with N basic modules; (b) deformed shape of MCPF.

Concerning a MCPF with N modules, if the one-sided output translation Δx is given, the deformation of each flexure can be calculated as

$$\delta x = \frac{\Delta x}{2N} \tag{2.10}$$

where $N \geq 1$. It is notable that $N = 1$ represents the case of traditional CPF.

It is observed from Fig. 2.3(a) that the N modules of a MCPF are connected in serial. Referring to Eq. (2.10) and Fig. 2.3(b), the stiffness of the MCPF can be computed as follows:

$$K^{\text{MCPF}} = \frac{F_x}{\Delta x} = \frac{2F}{2N\delta x} = \frac{Ebh^3}{Nl^3}. \tag{2.11}$$

Similarly, in consideration of Eqs. (2.4), (2.11), and (2.8), the maximum one-sided translation of the MCPF can be derived:

$$\Delta x_{\text{max}}^{\text{MCPF}} = \frac{F_{\text{max}}}{K^{\text{MCPF}}} = \frac{N\sigma_{\text{max}}l^2}{3Eh}. \tag{2.12}$$

By comparing Eq. (2.12) with Eq. (2.9), it can be deduced that the maximum translation of the MCPF is enlarged N times with respect to the ordinary CPF using flexures with the same material (σ_{max}/E) and physical (l and h) parameters.

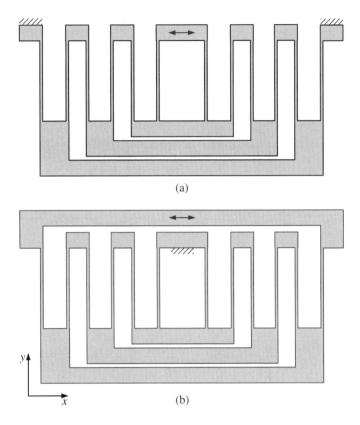

(a)

(b)

Figure 2.4   A MCPF with three basic modules ($N = 3$).

2.2   Design of a Large-Range Flexure Stage

2.2.1   Mechanism Design

Based on the idea of MCPF, a uniaxial flexure stage can be designed to achieve a large pure translational motion. For illustration, a MCPF is implemented with three basic modules by adopting $N = 3$, as shown in Fig. 2.4. In order to reduce the thermal expansion effect, a planar-symmetric or axial-symmetric structure design is preferred. To generate a symmetric structure, two identical MCPFs are employed to devise a compliant stage. According to different fixing schemes, four compliant stages are devised as shown in Fig. 2.5.

In addition, to increase the transverse stiffness of the MCPF in the y-axis, non working direction, the two secondary stages, which undergo identical displacement, can be connected using in-plane or out-of-plane links. For example, by employing in-plane and out-of-plane connections, the improved version of MCPF with $N = 3$ is designed as shown in Fig. 2.6(a) and (b) and Fig. 2.6(c) and (d), respectively. Using the improved MCPF, the enhanced uniaxial compliant stages are devised as shown in Fig. 2.7. The former two schemes in Fig. 2.7(a) and (b) are constructed using out-of-plane links and the latter ones in Fig. 2.7(c) and (d) are generated using in-plane connecting links.

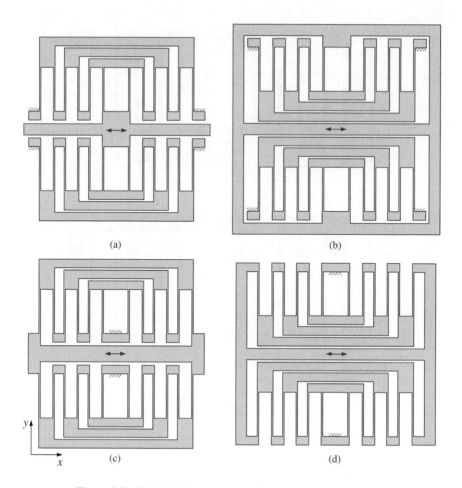

Figure 2.5   Uniaxial flexure stages with different fixing schemes.

Owing to the symmetric structure design of the uniaxial compliant stage, the parasitic motion is negligible in the xy-plane. For the purpose of demonstration, the scheme in Fig. 2.7(a) is chosen to design a micropositioning stage [5], as shown in Fig. 2.8. Analytical models are developed in the following sections to evaluate the stage performance.

2.2.2   Analytical Modeling

According to Eq. (2.12), the maximum two-sided output motion of the devised stage can be determined as

$$x_{max} = \frac{2\sigma_y l^2}{Eh}. \tag{2.13}$$

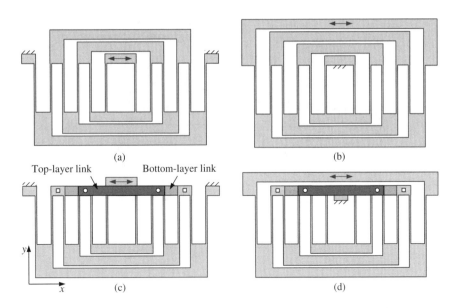

Figure 2.6 MCPF ($N = 3$) with improved transverse stiffness in y-axis: (a) and (b) are constructed using in-plane links; (c) and (d) are constructed using out-of-plane links.

The natural frequency of the stage is calculated as follows. First, the stiffness of a leaf flexure is computed. It has been shown that the stiffness can be determined by the bending deformation analysis, as shown in Eqs. (2.1) and (2.2). Alternatively, it can also be calculated by resorting to the pseudo-rigid-body (PRB) model [3]:

$$K = 2\gamma K_\theta \frac{EI}{l} \tag{2.14}$$

where γ is the optimal value of the characteristic radius, K_θ is a stiffness coefficient, and I is the area moment of inertia as shown in Eq. (2.3). In addition, the typical value of $\gamma = 0.85$ is adopted, and $K_\theta = 2.669$ is calculated by a linear interpolation according to the parameters given in the book [3].

The rotation angle of each leaf flexure is calculated as

$$\theta_z = \frac{x}{6} \times \frac{1}{l} \tag{2.15}$$

which is derived in view of Eq. (2.10). Thus, the potential energy is determined as

$$V = 48 \times \frac{1}{2} K \theta_z^2 = \frac{24 K x^2}{(6l)^2}. \tag{2.16}$$

Second, the kinetic energy of the stage is calculated as

$$T = \frac{1}{2} M \dot{x}^2 \tag{2.17}$$

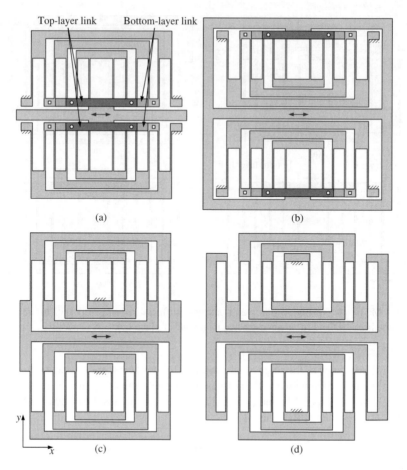

Figure 2.7   Different fixing schemes of the uniaxial flexure stages with improved transverse stiffness in the y-axis.

where x is the output displacement of the stage in the working direction and M is the equivalent mass, which is expressed as

$$M = m_a + 2m_b \left(\frac{5}{6}\right)^2 + 2m_c \left(\frac{3}{6}\right)^2 + 2m_d \left(\frac{1}{6}\right)^2 + 4m_e \left(\frac{4}{6}\right)^2$$

$$+ 4m_f \left(\frac{2}{6}\right)^2 + 4m_g \left(\frac{1 + \frac{5}{6}}{2}\right)^2 + 4m_g \left(\frac{\frac{5}{6} + \frac{4}{6}}{2}\right)^2$$

$$+ 4m_g \left(\frac{\frac{4}{6} + \frac{3}{6}}{2}\right)^2 + 4m_g \left(\frac{\frac{3}{6} + \frac{2}{6}}{2}\right)^2 + 4m_g \left(\frac{\frac{2}{6} + \frac{1}{6}}{2}\right)^2$$

$$+ 4m_g \left(\frac{\frac{1}{6} + 0}{2}\right)^2 + \frac{48}{(6l)^2} \times \frac{m_g l^2}{24} \tag{2.18}$$

where the mass components a, b, c, d, e, f, g are depicted in Fig. 2.8(b).

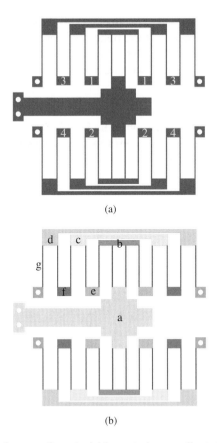

(a)

(b)

Figure 2.8 (a) Schematic diagram of a uniaxial long-stroke compliant stage; (b) illustration of mass components of the stage.

Third, substituting Eqs. (2.17) and (2.16) into the Lagrange's equation

$$\frac{d}{dt}\left(\frac{\partial T}{\partial \dot{x}}\right) - \frac{\partial T}{\partial x} + \frac{\partial V}{\partial x} = F \tag{2.19}$$

allows the generation of the free-motion dynamics equation (with $F = 0$):

$$M\ddot{x} + \frac{4K}{3l^2}x = 0 \tag{2.20}$$

which gives the natural frequency of the stage in units of hertz:

$$f = \frac{1}{2\pi}\sqrt{\frac{4K}{3l^2 M}}. \tag{2.21}$$

2.2.3 Architecture Optimization

Considering that a high natural frequency enables a large bandwidth of the servo system, the goal of the optimization is to achieve a natural frequency as large as possible under the constraints of large motion range and so on. Specifically, the stage architecture parameters are optimized to achieve the maximum natural frequency which is expressed by Eq. (2.21), and three constraints are taken into account during the optimization.

The first constraint comes from the motion range, i.e.,

$$x_{max} \geq 10 \, mm \tag{2.22}$$

where the two-sided stroke x_{max} is determined by Eq. (2.13).

The second constraint arises from the stiffness requirement of the stage. Due to the large motion-range requirement, the stage will be driven by a VCM. Given a VCM with the maximum actuation force ($F_{VCM} = 194.6$ N), the stage should be sufficiently compliant so that the VCM is able to drive it. The overall stiffness of the stage can be calculated below by substituting $N = 3$ into Eq. (2.11) along with consideration of the symmetry of the structure:

$$K_{stage} = \frac{2Ebh^3}{3l^3} \tag{2.23}$$

which is derived by taking into account that the two MCPFs are connected in parallel. Hence, the constraint is applied as

$$F_{VCM} \geq \frac{Ebh^3 x_{max}}{3l^3}. \tag{2.24}$$

The third constraint is imposed by the requirement of compact physical size (less than 100 mm × 100 mm) of the stage, as well as the manufacturing tolerance:

$$1 \, mm \leq l \leq 23 \, mm \tag{2.25}$$

$$0.5 \, mm \leq h \leq 0.9 \, mm. \tag{2.26}$$

In addition, the thickness of the material is selected as $b = 10$ mm.

The purpose of the optimization is to find the optimal parameters h and l to maximize the natural frequency as represented by Eq. (2.21). Due to the simple objective, analytical solutions can be found. However, the computation is complicated owing to the nonlinear functions with multiple variables and nonlinear constraints. In order to simplify the search procedure, the natural frequency is calculated as the changes in the two parameters h and l. The frequencies and parameter values, which satisfy the constraint conditions (2.22), (2.24), (2.25), and (2.26), are illustrated in Fig. 2.9. It is observed that the maximum natural frequency of 82.66 Hz is generated by using the parameters $h = 0.74$ mm and $l = 23$ mm. In addition, it is found that the maximum frequency is constrained by the condition (2.22), which is reflected by the generated stroke of 10 mm as shown in Table 2.1. The optimal parameters lead to a stage structure with an overall dimension within 100 mm × 100 mm. It is notable that numerical search algorithms can also be employed to find the optimal parameters.

To examine the performance of the optimized stage, FEA simulation is carried out with the ANSYS software package. Simulation results show that the stage has a natural frequency of 91.97 Hz along with a maximum one-sized stroke of 11.25 mm, which satisfies the performance requirements. In addition, the analytical and simulation results are tabulated in Table 2.1. The model error is calculated by taking the FEA simulation result as the benchmark. It is found that the analytical results are in good agreement with the simulation results, which verifies the accuracy of the established quantitative models of the stage.

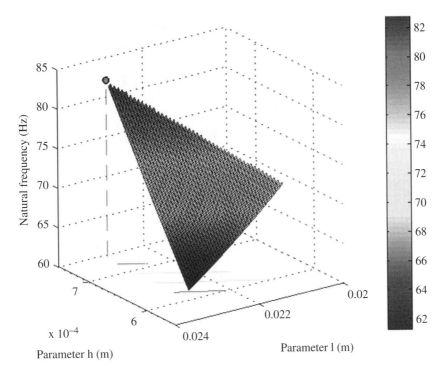

Figure 2.9   Natural frequency versus parameters h and l. The maximum frequency is marked by a circle.

Table 2.1   Main performances of the stage evaluated by different methods

Performance	Analytical model	FEA	Model error (%)
Natural frequency (Hz)	82.66	91.97	10.1
Maximum stroke (mm)	10.00	11.25	11.1

2.2.4   Structure Improvement

To assess the dynamics performance of the stage, modal analysis is also conducted with ANSYS software. As shown in Fig. 2.10(a), the first resonant mode shape is associated with the major motion (i.e., the translation along the working direction) and the second one is the rotation of the output stage in the xy-plane. Moreover, the first six resonant mode frequencies are shown in Table 2.2.

It is observed that the first two mode frequencies (i.e., 91.97 and 126.61 Hz), of the optimized stage are very close to each other, which indicates that the stage is prone to a rotational in-plane motion under external forces/torques due to a low rotational stiffness. Considering that this rotational motion is passive and cannot be controlled by the driving motor, it poses an obstacle to the control system design. Therefore, an improved structure design is necessary to enhance the passive rotational stiffness so as to improve the robustness of the stage motion along the major motion direction.

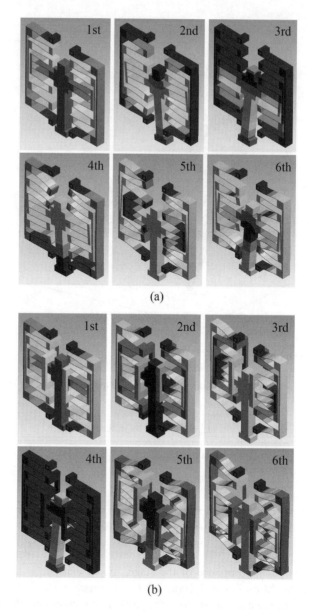

Figure 2.10   The first six mode shapes of (a) the original design and (b) the improved design.

An inspection of the motion property of the stage indicates that the two passive moving stages, which are denoted by the same numbers as shown in Fig. 2.8(a), translate displacements of identical magnitude. Hence, they can be connected together with connecting links, respectively, without influencing the motion property of the stage.

The first six mode shapes of the improved design are illustrated in Fig. 2.10(b). It is found that the first mode is still the translation along the working axis, whereas the original second

Table 2.2   First six mode frequencies of the optimized and improved stages

Mode no.	Optimal design (Hz)	Improved design (Hz)	Improvement (%)
1	91.97	102.34	11.3
2	126.61	226.71	79.1
3	161.22	244.32	51.5
4	169.77	261.69	71.8
5	269.04	339.22	26.1
6	291.26	364.65	25.2

mode as shown in Fig. 2.10(a) is eliminated. The second mode shape of the improved design is associated with the translation of the secondary stages. This indicates that the rotational stiffness of the improved design has been significantly enhanced compared with the original optimal design.

Moreover, the first six mode frequencies of the improved design are given in Table 2.2. It is observed that the first mode frequency is improved by 11.3%, even though the connecting bridges add the mass of the moving components. All of the other five mode frequencies have been improved by over 25%. In particular, for the improved design, the second mode frequency is substantially increased by 121.5% compared with the original design. Hence, the second mode frequency is over twice as high as the first one, which reveals a robust translational motion along the working direction. Therefore, from the control point of view, the improved design is more feasible than the original design.

2.3   Prototype Development and Performance Testings

With the optimized stage parameters, a prototype of the uniaxial micropositioning stage is developed as shown in Fig. 2.11. It is notable that the two sets of connecting links are mounted on the top and bottom of the stage, respectively. The stage has been fabricated by the wire-electrical discharge machining (EDM) process. To cater for the requirement on motion range, a VCM (model: NCC05-18-060-2X, from H2W Techniques, Inc.), which delivers a

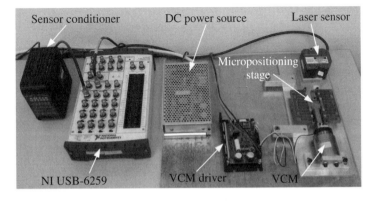

Figure 2.11   Experimental setup of the uniaxial micropositioning system.

stroke of 12.7 mm, is adopted to drive the stage. The output position of the stage is measured by a non-contact laser displacement sensor (model: LK-H055, from Keyence Corp.) with a measurement range of 20 mm. The laser displacement sensor provides an analog output signal and a digital output signal. Acquired by a data acquisition device, the analog output offers a displacement resolution around 1 μm. The digital output of the laser sensor is acquired by a computer through a USB or RS-232 interface, which provides the maximum sampling rate of 392 kHz. The digital output delivers a finer resolution of sub-micrometer level. Moreover, an NI USB-6259 board (from National Instruments Corp.) equipped with 16-bit A/D and D/A channels is used to implement the controller hardware. The control algorithms are developed with LabVIEW software.

2.3.1 Statics Performance Testing

By applying a 0.5-Hz low-frequency sinusoidal voltage signal to drive the VCM, the stage output position is measured by the laser displacement sensor. With a conservative magnitude of ±2.3 V for the input voltage, a motion range of 11.25 mm is obtained, which is larger than the desired value of 10 mm. Thus, the experimental tests verified the large motion range of the developed micropositioning stage. Evidently, the design objective is well achieved.

In addition, as the input frequency increases, the output–input relationship curves are obtained as shown in Fig. 2.12, which exhibit clear hysteresis phenomena. The hysteresis effects mainly come from the dynamics of the flexure stage in conjunction with the adopted VCM actuator. Hysteretic nonlinearity has been well recognized for systems driven by smart material-based actuators, such as piezoelectric actuators and shape memory alloy actuators, etc. The presence of a hysteresis effect necessitates the development of control techniques to achieve precision positioning.

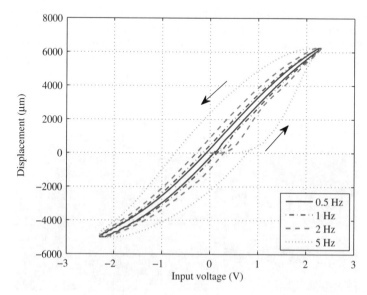

Figure 2.12   Motion range test results of the micropositioning stage with different frequencies of sinusoidal input signals.

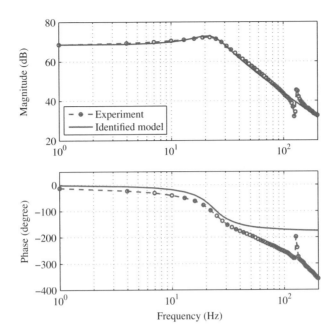

Figure 2.13 Open-loop frequency responses of the system obtained by experimental results and the identified model.

2.3.2 Dynamics Performance Testing

The dynamics performance of the micropositioning system is tested by the frequency response method. Specifically, the swept-sine waves with amplitude of 0.05 V and frequency range of 1–1000 Hz are produced by a D/A channel of the NI USB-6259 board to drive the VCM. The input–output data sets are used to generate the frequency response of the stage. From the result as shown in Fig. 2.13, a resonant frequency of 23.4 Hz can be identified. Compared with the FEA result, the lower resonant frequency of the system mainly comes from the mass of the moving coil of the motor, which is not considered in the FEA simulation, and the fabrication error of the stage.

2.4 Sliding Mode Controller Design

In order to implement a precise positioning control of the micropositioning system, a robust DSMC control scheme is developed and tested.

2.4.1 Dynamics Modeling

The mathematical models of VCM have been well studied before [6]. In the current work, the linear VCM is driven by a VCM driver, which converts an input voltage into a current for the drives. The whole system (including VCM and flexure stage) is modeled as a mass–spring–damper system as follows:

$$\hat{M}\ddot{x}(t) + \hat{B}\dot{x}(t) + \hat{K}x(t) = \hat{D}u(t) + P(t) \tag{2.27}$$

where t is the time variable and parameters $\hat{M}$, $\hat{B}$, $\hat{K}$, and $\hat{D}$ represent the estimated mass, damping coefficient, stiffness, and force constant of the system, respectively. In addition, u denotes the input voltage applied to the VCM driver, x is the output displacement of the stage, and $P(t)$ is the perturbation term which describes the lumped effect of the hysteresis, model parameter uncertainties, unmodeled high-frequency dynamics, and other disturbances.

First, dividing both sides of Eq. (2.27) by $\hat{M}$ results in

$$m\ddot{x}(t) + b\dot{x}(t) + kx(t) = du(t) + p(t) \tag{2.28}$$

where $m = 1$, $b = \frac{\hat{B}}{\hat{M}}$, $k = \frac{\hat{K}}{\hat{M}}$, $d = \frac{\hat{D}}{\hat{M}}$, and $p(t) = \frac{P(t)}{\hat{M}}$.

By adopting a small sampling time T, the following approximations are valid [7]:

$$\dot{x}(t) \approx \frac{1}{T}(x_k - x_{k-1}) \tag{2.29}$$

$$\ddot{x}(t) \approx \frac{1}{T^2}(x_k - 2x_{k-1} + x_{k-2}) \tag{2.30}$$

where $x_k = x(kT)$ with k representing the kth time step.

Then, the continuous-time dynamics model in Eq. (2.28) can be converted into an equivalent discrete-time form:

$$\bar{m}x_{k-2} + \bar{b}x_{k-1} + \bar{k}x_k = \bar{d}u_k + p_k \tag{2.31}$$

where

$$\bar{m} = \frac{1}{T^2}, \quad \bar{b} = -\frac{b}{T} - \frac{2}{T^2}, \quad \bar{k} = k + \frac{b}{T} + \frac{1}{T^2}, \quad \bar{d} = d. \tag{2.32}$$

Based on the perturbation estimation technique [8], the perturbation term p_k can be generated by its one-step delayed estimation

$$\hat{p}_k = p_{k-1} = -\bar{d}u_{k-1} + \bar{m}x_{k-3} + \bar{b}x_{k-2} + \bar{k}x_{k-1}. \tag{2.33}$$

Hence, the dynamics model (2.31) can be rewritten as

$$\bar{m}x_{k-2} + \bar{b}x_{k-1} + \bar{k}x_k = \bar{d}u_k + \hat{p}_k - \tilde{p}_k \tag{2.34}$$

where $\tilde{p}_k = \hat{p}_k - p_k$ is the perturbation estimation error.

The concerned control problem is how to achieve a precise position control in the presence of estimation error $\tilde{p}_k$. To achieve this objective, a DSMC controller design is presented in the following discussion.

2.4.2 DSMC Design

Before the control design, the position error is defined as follows:

$$e_k = x_k - r_k \tag{2.35}$$

where x_k and r_k are the actual and reference position outputs at the kth time step, respectively.

Based on the position error e_k, an incremental proportional-integral-derivative (PID) type of sliding function is defined as follows:

$$s_k = K_p(e_k - e_{k-1}) + \frac{K_pT}{T_i}e_k + \frac{K_pT_d}{T}(e_k - 2e_{k-1} + e_{k-2}) \tag{2.36}$$

where T is the sampling time. K_p, T_i, and T_d denote the proportional gain, integral time, and derivative time, respectively.

After a necessary algebra operation, the sliding function (2.36) is simplified as

$$s_k = K_3e_{k-2} + K_2e_{k-1} + K_1e_k \tag{2.37}$$

where

$$K_1 = K_p + \frac{K_pT}{T_i} + \frac{K_pT_d}{T} \tag{2.38}$$

$$K_2 = -K_p - \frac{2K_pT_d}{T} \tag{2.39}$$

$$K_3 = \frac{K_pT_d}{T}. \tag{2.40}$$

Regarding the system as described by Eq. (2.34) with the sliding function (2.37), a DSMC controller is devised as follows:

$$u_k = u_{k-1} + \frac{\bar{k}}{\bar{d}K_1}(s_{k-1} - K_3e_{k-2} - K_2e_{k-1} + K_1r_k)$$

$$-\frac{1}{\bar{d}}[\bar{m}x_{k-3} + (\bar{b} - \bar{m})x_{k-2} + (\bar{k} - \bar{b})x_{k-1}] - \frac{K_s}{\bar{d}}\text{sign}(s_{k-1}) \tag{2.41}$$

where sign($\cdot$) denotes the signum function and K_s is a positive switching gain.

The following stability analysis reveals that the discrete sliding mode will occur after a finite number of steps.

First, substituting the estimated perturbation term in Eq. (2.33) into the dynamic model in Eq. (2.34) allows the calculation of the position:

$$x_k = \frac{1}{\bar{k}}[\bar{d}(u_k - u_{k-1}) + \bar{m}x_{k-3} + (\bar{b} - \bar{m})x_{k-2} + (\bar{k} - \bar{b})x_{k-1} - \tilde{p}_k]. \tag{2.42}$$

Then, substituting Eqs. (2.41) and (2.42) into the sliding function Eq. (2.37), a necessary algebra manipulation leads to

$$s_k = K_3e_{k-2} + K_2e_{k-1} + K_1(x_k - r_k)$$

$$= s_{k-1} - \frac{K_1}{\bar{k}}\tilde{p}_k - \frac{K_1K_s}{\bar{k}}\text{sign}(s_{k-1})$$

$$= s_{k-1} - \frac{K_1}{\bar{k}}[\tilde{p}_k + K_s\text{sign}(s_{k-1})] \tag{2.43}$$

where the parameters $\bar{k}$ and K_1 are positive definite.

In the case of $s_{k-1} \geq 0$, it can be deduced that, if $K_s \geq |\tilde{p}_k|$ holds, then

$$s_k = s_{k-1} - \frac{K_1}{\bar{k}}(\tilde{p}_k + K_s)$$

$$\leq s_{k-1}. \tag{2.44}$$

Otherwise, in the case of $s_{k-1} < 0$, the following deduction can be derived:

$$s_k = s_{k-1} - \frac{K_1}{\bar{k}}(\tilde{p}_k - K_s)$$

$$\geq s_{k-1} \tag{2.45}$$

provided that $K_s \geq |\tilde{p}_k|$ holds.

Hence, in consideration of Eqs. (2.44) and (2.45), the following conclusion can be drawn:

$$|s_k| \leq |s_{k-1}| \quad \text{if} \quad K_s \geq |\tilde{p}_k| \tag{2.46}$$

which indicates that s_k decreases monotonously and the discrete sliding mode is reached after a finite number of steps.

The relationship in Eq. (2.46) represents a sufficient condition for the existence of a discrete sliding mode [9]. Due to the discontinuity of the signum function $\text{sign}(\cdot)$, chattering may occur in the control input. To alleviate the chattering phenomenon, the boundary layer technique can be adopted by replacing the signum function in Eq. (2.41) with the saturation function

$$\text{sat}(s_k) = \begin{cases} \text{sign}(s_k) & \text{for} \quad |s_k| > \epsilon \\ \dfrac{s_k}{\epsilon} & \text{for} \quad |s_k| \leq \epsilon \end{cases} \tag{2.47}$$

where the parameter ϵ denotes the boundary layer thickness of the sliding function s_k. It is notable that use of the saturation function reduces the chattering effect at the cost of a robustness reduction. In practice, a tradeoff between the chattering effect and tracking error is needed to assign the parameter ϵ.

2.5 Experimental Studies

In this section, several experimental studies are carried out to achieve precision positioning for the micropositioning system.

2.5.1 Plant Model Identification

Based on the frequency response of the system as shown in Fig. 2.13, a second-order model can be identified for the system plant as follows:

$$G(s) = \frac{5.839 \times 10^7}{s^2 + 94.27s + 2.17 \times 10^4}. \tag{2.48}$$

It is observed from Fig. 2.13 that the second-order model is capable of approximating the dynamics behavior up to 100 Hz. In order to capture the system dynamics at higher frequencies, a model of much higher order is required to be identified. In this work, a simple second-order model is employed to demonstrate the effectiveness of the developed DSMC control scheme.

By comparing Eq. (2.28) and the inverse Laplace transform of Eq. (2.48), the dynamics model parameters can be derived as: $b = 94.27$ N·s/μm, $k = 2.17 \times 10^4$ N/μm, and $d = 5.839 \times 10^7$ μm/V. Using a sampling time of $T = 0.005$ s, the discrete-time model parameters can be obtained from Eq. (2.32).

2.5.2 Controller Setup

For the purpose of comparison, PID control is employed due to its popularity [10]. It is known that PID is a model-free controller, which solves the control command by making use of the tracking error only. In practice, a digital PID control algorithm can be realized as follows:

$$u_k = K_p e_k + \frac{K_p T}{T_i} \sum_{j=0}^{k} e_j + \frac{K_p T_d}{T}(e_k - e_{k-1}) \tag{2.49}$$

where the positioning error $e_k = r_k - x_k$. In addition, K_p, T_i, and T_d denote the proportional gain, integral time constant, and derivative time constant, respectively.

The key issue of PID design is the tuning of the three control parameters (K_p, T_i, and T_d). In this work, the Ziegler–Nichols (Z–N) method is adopted to determine the parameters. Specifically, by setting $T_i = \infty$ and $T_d = 0$ and increasing K_p from zero, the ultimate gain $K_u = 2.06 \times 10^{-4}$ produces a critical oscillation with a constant amplitude and an oscillation period $T_u = 0.04$ s. The classic Z–N method suggests PID parameters as follows: $K_p = 0.6K_u$, $T_i = 0.5T_u$, and $T_d = 0.125T_u$.

For a comparison study, the tuned PID gains are adopted to determine the control gains K_1, K_2, and K_3 of the DSMC scheme by using Eqs. (2.38), (2.39), and (2.40), respectively. Compared with the PID control algorithm, the presented DSMC has an additional switching gain K_s, which is introduced to suppress the perturbation estimation error $\tilde{p}_k$. In practice, the switching gain K_s and the boundary layer thickness ϵ are adjusted by a trial-and-error approach through experimental studies.

In the following discussion, experimental studies are conducted to verify the performance of the micropositioning system with the developed control schemes.

2.5.3 Set-Point Positioning Results

For illustration, a 1-mm set-point positioning is carried out. The PID gains are tuned by using the classic Z–N method, which produces a positioning result as shown in Fig. 2.14. Concerning the DSMC algorithm, the boundary layer parameter is assigned as $\epsilon = 2.0$, and the positioning results of different switching gains K_s are shown in Fig. 2.14. It is seen that the DSMC achieves a much quicker transient response than the PID control. In addition, the phase portrait of the DSMC is depicted in Fig. 2.15. It is found that both the sliding function s_k and its change rate Δs_k converge to the origin, which illustrates the stability of the system graphically.

To evaluate the reliability of the micropositioning system with the developed controllers, the set-point positioning is repeated five times using each controller. The DSMC controllers 1 to 4 are designed with switching gains K_s of 1×10^7, 3×10^7, 5×10^7, and 7×10^7, respectively. Figure 2.16 shows the error bars of the three main performances, including the steady-state root-mean-square errors (RMSEs), 5% settling time, and percentage overshoot.

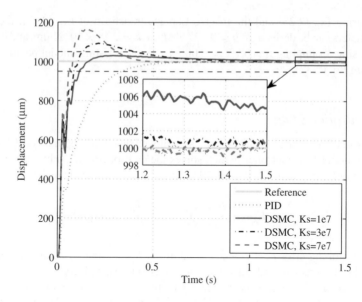

Figure 2.14 Set-point positioning results of PID and DSMC with different switching gain K_s.

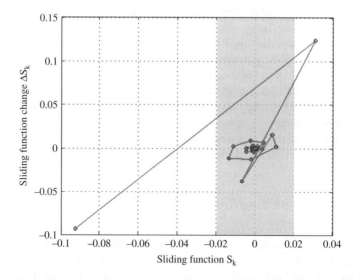

Figure 2.15 Phase portrait of the DSMC with switching gain $K_s = 10^7$.

It is observed that the PID control produces an average steady-state RMSE of 0.62 μm with a 5% settling time of 0.33 s. As the switching gain K_s increases, the produced RMSE and overshoot by the DSMC scheme gradually decreases and increases, respectively. However, the variation tendency of the settling time is not monotonous. The reason is that as the switching gain increases, the settling time is determined by either the lower or the upper 5%

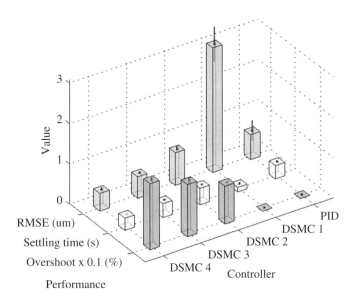

Figure 2.16   Error bars of positioning RMSE, 5% settling time, and overshoot of the positioning system with different controllers.

error boundary as shown in Fig. 2.14. In addition, the PID and DSMC with a smaller gain $K_s = 1 \times 10^7$ produce zero overshoot. Compared with PID, the DSMC with a smaller switching gain leads to a smaller settling time of 0.11 s, which is obtained at the cost of a larger steady-state RMSE of 3.09 μm. In contrast, a larger switching gain $K_s = 7 \times 10^7$ enables a smaller RMSE of 0.44 μm and a shorter settling time of 0.31 s at the cost of a larger overshoot of 16.43% for the DSMC scheme.

By adopting DSMC with a large switching gain 7×10^7, the resolution of the micropositioning system is tested by a consecutive set-point positioning testing as shown in Fig. 2.17. It is observed that the step size of 250 nm can be clearly identified, which demonstrates a resolution better than 250 nm for the micropositioning system. It is notable that the positioning resolution can be further enhanced by employing a displacement sensor with finer resolution and lower noise level.

2.5.4   Sinusoidal Positioning Results

The positioning performance of the micropositioning system for a sinusoidal motion as shown in Fig. 2.18(a) is verified. To accomplish a better positioning accuracy, the PID gains are tuned by employing the Pessen integral rule, which produces the following PID gains: $K_p = 0.7K_u$, $T_i = 0.4T_u$, and $T_d = 0.15T_u$.

Compared with the classic Z–N method, the Pessen integral rule gives a larger proportional control action, which leads to an explicit oscillation in the set-point positioning in the current work. In contrast, the PID gains tuned by the Pessen integral rule result in a much smaller error for the sinusoidal positioning, as shown in Fig. 2.18(b). It is observed that a maximal error of 24.43 μm is obtained, which accounts for a relative error of 2.44% relative to the motion range.

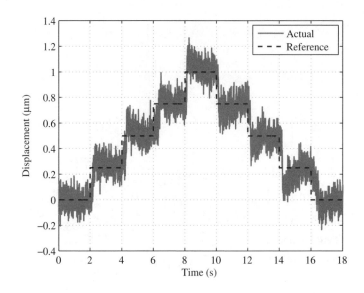

Figure 2.17   Consecutive 250-nm set-point positioning result of the system.

It is notable that this PID result is much better than the result of 3.74% obtained by a classic Z–N tuned PID.

Using the aforementioned PID gains, other parameters of DSMC are adjusted as $K_s = 7 \times 10^7$ and $\epsilon = 0.2$. The positioning result of the DSMC is also depicted in Fig. 2.18, which reveals a maximal positioning error of 5.31 μm (i.e., 0.53% relative error). Compared with PID, the DSMC has substantially reduced the maximal positioning error by 78%. Evidently, the positioning results demonstrate the superiority of the DSMC strategy over the PID algorithm.

It is notable that the gain parameters K_1, K_2, and K_3 of the presented DSMC controller are determined directly by the tuned PID gains; the switching gain K_s and boundary layer parameter ϵ are adjusted by trial-and-error in experimental studies. That is, the control gains are not optimally designed in the preceding experiments. The current amplitudes of the positioning errors indicate that the performance of the micropositioning system with the DSMC scheme may be improved to further improve the positioning accuracy. In the future, displacement sensors with finer resolution (e.g., laser interferometer) will be employed to enhance the system positioning performance. Additionally, the presented approaches can also be extended to the design and control of multi-axis large-range micropositioning systems.

2.6   Conclusion

The mechanism and control designs of a uniaxial large-range micropositioning system are presented in this chapter. The micropositioning stage is composed of multi-stage compound parallelogram flexures, which enable a sub-micrometer positioning resolution (250 nm), a centimeter-level long stroke (over 11 mm), and compact dimension (within 100 mm × 100 mm) simultaneously. The architecture parameters of the stage are optimized to generate the highest natural frequency under the constraints on motion range and physical size. The long stroke is confirmed by both FEA and experimental results. In addition, a DSMC algorithm is devised to achieve precision positioning in the presence of model uncertainties and disturbances. The effectiveness of the developed control scheme is validated by experimental studies along with a comparative study with the popular PID algorithm. The uniaxial large-range flexure stage can be employed in various applications, such as the auto-focusing of a microscope as demonstrated in the recent work [11].

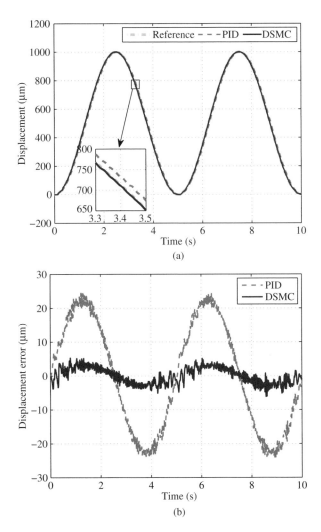

Figure 2.18 Sinusoidal positioning results of the micropositioning system with PID and DSMC control schemes.

References

[1] Cannon, B.R., Lillian, T.D., Magleby, S.P., Howell, L.L., and Linford, M.R. (2005) A compliant end-effector for microscribing. *Precis. Eng.*, **29** (1), 86–94.

[2] Teo, T.J., Yang, G., and Chen, I.M. (2014) A large deflection and high payload flexure-based parallel manipulator for UV nanoimprint lithography: Part I. Modeling and analysis. *Precis. Eng.*, **38** (4), 861–871.

[3] Howell, L.L. (2001) *Compliant Mechanisms*, Wiley, New York.

[4] Xu, Q. (2012) New flexure parallel-kinematic micropositioning system with large workspace. *IEEE Trans. Robot.*, **28** (2), 478–491.

[5] Xu, Q. (2013) Design, testing and precision control of a novel long-stroke flexure micropositioning system. *Mech. Mach. Theory*, **70**, 209–224.

[6] Ratliff, R.T. and Pagilla, P.R. (2005) Design, modeling, and seek control of a voice-coil motor actuator with nonlinear magnetic bias. *IEEE Trans. Magn.*, **41** (6), 2180–2188.

[7] Tarokh, M. (1990) A discrete-time adaptive control scheme for robot manipulators. *J. Robotic Systems*, **7** (2), 145–166.

[8] Elmali, H. and Olgac, N. (1996) Implementation of sliding mode control with perturbation estimation (SMCPE). *IEEE Trans. Control Syst. Technol.*, **4** (1), 79–85.

[9] Sarpturk, S., Istefanopulos, Y., and Kaynak, O. (1987) On the stability of discrete-time sliding mode control systems. *IEEE Trans. Automat. Contr.*, **32** (10), 930–932.

[10] Dong, W., Tang, J., and ElDeeb, Y. (2009) Design of a linear-motion dual-stage actuation system for precision control. *Smart Mater. Struct.*, **18** (9), 095–035.

[11] Liu, Y. and Xu, Q. (2015) Design of a flexure-based auto-focusing device for a microscope. *Int. J. Precis. Eng. Manuf.*, **16** (11), 2271–2279.

3

XY Flexure Stage

Abstract: This chapter presents the design and development of a decoupled parallel-kinematic XY compliant micropositioning system which has a motion range larger than 10 mm in each axis along with a compact structure. Quantitative models are established to predict the stage performances, which are validated by conducting finite-element analysis (FEA) and experimental studies. The scheme of model predictive control (MPC) is implemented for precision motion control of the system, which possesses a non-minimum-phase plant. The MPC is capable of producing a low magnitude of output tracking error by imposing an appropriate suppression on the control effort. The superiority of MPC over proportional-integral-derivative (PID) control is demonstrated by both simulation and experimental investigations.

Keywords: Two-dimensional micropositioning, XY flexure mechanisms, Parallel mechanisms, Translational stages, Total decoupling design, Finite-element analysis, Model predictive control, Motion control, Non-minimum-phase system.

3.1 Introduction

To realize precision in-plane translational micropositioning, several XY flexure stages with large motion range have been proposed in the literature. For example, a compact stage is presented in [1], which has a dimension of 100 mm $\times$ 100 mm $\times$ 50 mm. However, it can only deliver a small workspace of 2 mm $\times$ 2 mm. In contrast, the XY stage designed in [2] produces a workspace of 10 mm $\times$ 10 mm, which is contributed by a large physical dimension of 255 mm $\times$ 255 mm $\times$ 25 mm. A compact stage is desirable for situations where the micro-/nanomanipulation is required to be executed inside a limited space (e.g., inside the chamber of a scanning electron microscope) [3]. However, it is challenging to design an XY stage with both a large workspace and a compact physical dimension simultaneously. In order to make a quantitative measure, the area ratio[1] is proposed here for the purpose of comparison study. A larger ratio implies that the stage can be fabricated as a smaller dimension for achieving a specified workspace size. Thus, the larger the ratio, the more compact the stage.

---

[1] The area ratio is defined as the ratio between the area of the planar workspace and the area of the planar dimension of the XY stage.

Design and Implementation of Large-Range Compliant Micropositioning Systems, First Edition. Qingsong Xu.
© 2016 John Wiley & Sons Singapore Pte Ltd. Published 2016 by John Wiley & Sons Singapore Pte Ltd.

In practice, it is difficult to design a flexure XY micropositioning stage with a large workspace range, say 10 mm × 10 mm, along with a compact dimension for micropositioning applications. In this chapter, the concept of MCPF as reported in Chapter 2 is employed to generate a large translational motion in each working direction. A totally decoupled large-workspace XY stage is designed and fabricated along with the unwanted buckling/bending phenomenon eliminated. The parallel-kinematic XY stage exhibits both input decoupling and output decoupling properties. However, the system identification results show that the plant model of the micropositioning system is of non-minimum-phase (NMP), which poses an obstacle for the control system design.

NMP indicates that the system model possesses unstable zeros. Therefore, many control techniques based on high-gain feedback and the inversion principle cannot be applied directly. Mathematically, the unstable zeros can be removed by a pole–zero cancellation approach. However, this is not feasible in practice because an arbitrarily small discrepancy between the zero and the pole causes instability [4]. On the contrary, a NMP system can be controlled by using a zero-phase-error tracking controller (ZPETC) or a zero-magnitude-error tracking controller (ZMETC) [5]. However, ZPETC creates a zero phase error at the cost of magnitude errors. Similarly, ZMETC accomplishes fine magnitude tracking as a sacrifice of phase delays. Hence, significant revisions are required to pursue perfect tracking [6]. An output feedback sliding mode control algorithm with sufficient revisions [7] is also applicable to NMP systems. Nevertheless, the implementation of such controllers is a complicated and time-consuming procedure. In contrast, it has been shown [8] that MPC provides an efficient method of control for the NMP system. As a robust control strategy, MPC is well-known for its ability to solve problems with constraints, time delays, and disturbances by offering an optimal control effort [9].

Generally, a MPC controller is easy to implement because it has a sole weight parameter to tune [10]. At the same time, fewer tunable parameters implies less flexibility in design. Preliminary study shows that the conventional MPC cannot produce a satisfactory positioning in terms of transient response time and steady-state error by adjusting the sole parameter. To overcome such limitation, an enhanced model predictive control (EMPC) is proposed by resorting to the popular PID controller. It is demonstrated that the EMPC derives an optimal incremental control action by imposing suitable suppression on the control effort. The theoretical analysis and effectiveness of the EMPC are verified by simulation and experimental studies performed on the micropositioning system with NMP plant.

3.2   XY Stage Design

3.2.1   Decoupled XY Stage Design with MCPF

In this work, the MCPF with $N = 2$ is selected as a demonstration to design an XY stage. Figure 3.1(a) and (b) shows two different fixing schemes of the MCPF. It is notable that $N = 1$ or $N \geq 3$ can also be used, and the design procedure is similar to the one presented below.

To design a decoupled XY stage, a 2-PP parallel mechanism[2] is employed as shown in Fig. 3.2(a). One embodiment of flexure realization of the 2-PP mechanism is shown in Fig. 3.2(b). The two sets of MCPFs, which are connected to the actuators, also act as

---

[2] P stands for a prismatic joint.

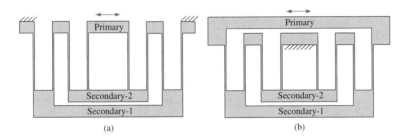

Figure 3.1   MCPF ($N = 2$) with different fixing schemes.

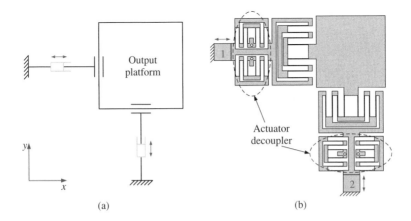

Figure 3.2   (a) A 2-PP parallel mechanism; (b) a flexure 2-PP parallel mechanism.

actuator decouplers. The other two sets of MCPFs, which are connected to the output platform, ensure decoupled output translations in the x- and y- axes. Hence, the flexure mechanism possesses the property of input decoupling and output decoupling (i.e., it exhibits the total decoupling property) [11]. The actuator decoupler exhibits a large transverse stiffness at the output end, which guarantees that the actuator does not suffer from transverse loading.

In order to generate a more compact XY stage with larger area ratio, the decoupler is evolved as shown in Fig. 3.3(a). In this way, a larger output platform is obtained and the planar space is utilized more efficiently. Alternative designs of the compact flexure 2-PP parallel mechanism are illustrated in Fig. 3.3(b)–(h), with different fixing schemes. Similarly, more variations of the flexure XY stage can be obtained.

Furthermore, to improve the accuracy performance and reduce the temperature-gradient effect, a mirror-symmetric XY stage can be obtained by adopting the 4-PP parallel mechanism, as shown in Fig. 3.4(a). For example, such a monolithic flexure XY stage is implemented as shown in Fig. 3.4(b). It is notable that each variation in Fig. 3.3 can be improved as a mirror-symmetric 4-PP flexure mechanism. Due to the symmetric structure, the in-plane parasitic motion of the XY stage can be neglected.

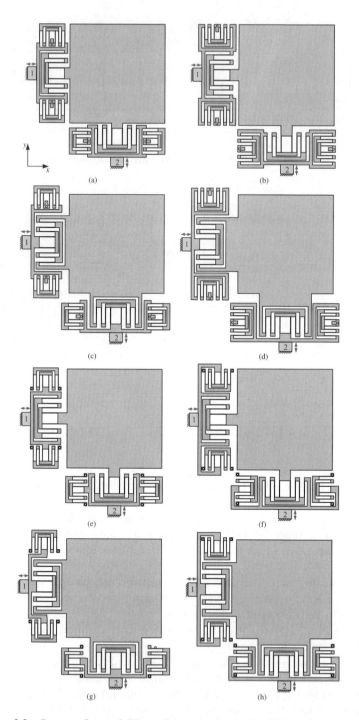

Figure 3.3   Compact flexure 2-PP parallel mechanisms with different fixing schemes.

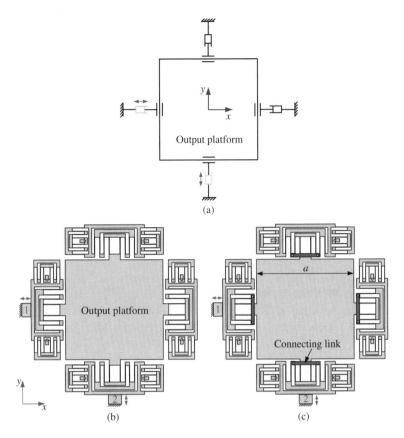

Figure 3.4 (a) A 4-PP parallel mechanism with two P joints actuated; (b) a monolithic, mirror-symmetric XY stage; (b) an improved XY stage with enhanced transverse stiffness.

3.2.2 Buckling/Bending Effect Consideration

Considering that the flexures are constructed by slender leaf springs with a relatively high ratio of length to thickness, they are prone to elastic (Euler) buckling/bending under compressive loads, as shown in Fig. 3.5(a). As an adverse effect, the motion that can be transmitted to the output platform is greatly reduced. Thus, the buckling/bending effect should be avoided by the structure design.

In theory, the critical axial load that causes elastic buckling of the flexure can be calculated as

$$P_{cr} = \frac{\pi^2 EI}{l_{cr}^2} \tag{3.1}$$

where the critical length l_{cr} is evaluated by using a specific coefficient k as follows:

$$l_{cr} = kl. \tag{3.2}$$

It is known that k takes values from 0.5 to 2 depending on the boundary conditions [12].

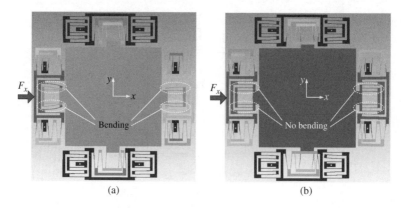

(a)   (b)

Figure 3.5   FEA results of deformations for an XY stage (a) without and (b) with the connecting links.

Referring to the FEA results shown in Fig. 3.5(a), assume that the MCPF deformation is induced by the buckling effect. The buckling flexures are considered as fixed–free beams, which have a coefficient of $k = 2$. Thus, the ordinary MCPF has the critical load value

$$P_{cr}^{MCPF1} = \frac{\pi^2 EI}{4l^2}. \tag{3.3}$$

In order to enlarge the critical load to avoid the occurrence of buckling, either the moment of inertia I can be increased or the critical length l_{cr} can be decreased, as revealed in Eq. (3.1). Regarding a flexure designed with specific dimensions, a method is proposed below to reduce the critical length l_{cr} by reducing the coefficient k.

By investigating the deformed shape of the two-stage MCPF ($N = 2$) as shown in Fig. 2.3(b), it is found that the displacements of the two free ends are identical, i.e., $x_2 = x_4 = \Delta x/2$. Hence, these two ends can be connected together without influencing the motion output property. Once the two free ends of each MCPF are connected by a connecting link as shown in Fig. 3.4(c), the concerned flexures become fixed–fixed columns. For a beam with fixed–fixed boundary condition, the coefficient is reduced to $k = 0.5$ [12]. Thus, the critical load can be computed as

$$P_{cr}^{MCPF2} = \frac{4\pi^2 EI}{l^2}. \tag{3.4}$$

It can be observed from Eqs. (3.4) and (3.3) that the critical load of the improved design has been enhanced 16 times in comparison with the original design. Even if the deformation as shown in Fig. 3.5(a) is merely due to the bending effect, the critical load will be improved further since the bending load is lower than the critical buckling load (3.3). Therefore, the buckling/bending phenomenon is alleviated by the improved XY stage as shown in Fig. 3.4(c). The FEA result is shown in Fig. 3.5(b). It is observed that, with the same force applied to one input end of the XY stage, no buckling/bending occurs in the related flexures. As a result, a larger output displacement is generated compared with the original XY stage as shown in Fig. 3.5(a).

3.2.3   Actuation Issues

As far as actuation is concerned, two VCMs are selected to drive the XY stage in this work to generate a centimeter-level motion range. In comparison with other types of actuators based on smart materials such as the piezoelectric stack actuator (PSA), VCM provides a larger stroke. However, VCM has a low blocking force compared with PSA. To facilitate VCM actuation, the XY stage should be designed with sufficiently low stiffness. At the same time, the actuation force should not induce a buckling phenomenon.

Quantitatively, to guarantee the elimination of the buckling effect for the XY stage, the maximum actuation force should not exceed the critical force of actuation:

$$F_{cr} = 2P_{cr}^{MCPF2} = \frac{8\pi^2 EI}{l^2}. \tag{3.5}$$

It can be observed from Fig. 3.5(b) that, when an actuation force F_x is applied on the XY stage to generate a displacement x of the output platform, the deformations are primarily induced by the bending of six MCPFs. Taking into account the bending deformations only, the stiffness model of the XY stage is simplified as shown in Fig. 3.6. The stiffness K of each MCPF can be computed by considering $N = 2$ in Eq. (2.11):

$$K = \frac{Ebh^3}{2l^3}. \tag{3.6}$$

Then, the stiffness of the XY stage observed at the input end can be computed as

$$K_{in} = 6K = \frac{3Ebh^3}{l^3}. \tag{3.7}$$

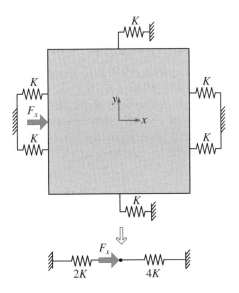

Figure 3.6   Stiffness model of the XY flexure stage.

Given the predefined maximum one-sided displacement D_{max} of the stage in one axial direction, the required driving force from the motor can be expressed as

$$F_{max} = K_{in}D_{max}. \tag{3.8}$$

When selecting the motors, the VCM should be chosen with the maximum actuation force F_{max}^{VCM} satisfying

$$F_{max} \leq F_{max}^{VCM} \leq F_{cr}. \tag{3.9}$$

That is,

$$\frac{3Ebh^3 D_{max}}{l^3} \leq F_{max}^{VCM} \leq \frac{2\pi^2 Ebh^3}{3l^2}. \tag{3.10}$$

The above condition guarantees that the selected motor is powerful enough to drive the XY stage, yet does not excite a buckling response of the structure at the same time.

In addition, the given maximum one-sided displacement D_{max} should stay within the allowable maximum one-sided translation described by Eq. (2.12) for $N = 2$, i.e.,

$$D_{max} \leq \frac{2\sigma_{max}l^2}{3Eh}. \tag{3.11}$$

Therefore, the relationships (3.10) and (3.11) provide guidelines for the design of stage parameters to guarantee good operation with safety of the material.

3.3   Model Verification and Prototype Development

In this work, an XY flexure stage is developed to produce a motion range over ± 5 mm in each working axis. Based on the design criteria (3.10) and (3.11), the XY stage is designed with the main parameters as shown in Table 3.1. The parameters are denoted in Figs. 2.2 and 3.4(c).

3.3.1   Performance Assessment with FEA Simulation

In the previous section, the quantitative models were established under certain assumptions. In order to verify the accuracy of the developed models and to evaluate the performance of the designed stage, FEA simulations are carried out with the ANSYS software package. The main specifications of the adopted material (Al 7075 alloy) are shown in Table 3.2.

The analytical models reveal that the maximum allowable one-sided displacement of the stage is $\Delta x_{max} = 5.85$ mm. Thus, the motion range in each working axis is $\pm\Delta x_{max} = \pm 5.85$

Table 3.1   Main structural parameters of the parallel-kinematic XY flexure stage

Parameter	Symbol	Value	Unit
Length of the MCPF flexure	l	25.0	mm
In-plane width of the MCPF flexure	h	0.5	mm
Out-of-plane thickness of the MCPF flexure	b	10.0	mm
Edge length of the output platform	a	122.0	mm

Table 3.2 Main parameters of the Al 7075 alloy material

Parameter	Value	Unit
Young's modulus (E)	71.7	GPa
Tensile yield strength (σ_y)	503	MPa
Poisson's ratio	0.33	—
Density	2810	kg/m^3

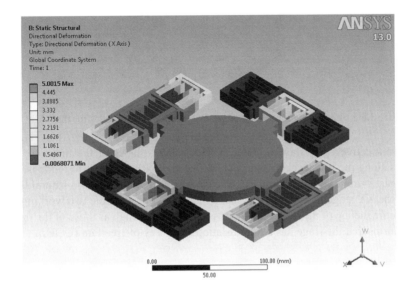

Figure 3.7 FEA simulation results of static deformations for the XY flexure stage.

mm. By assigning a motion range of ±5 mm, the required maximum force to drive the stage is derived as $F_{max} = 86.04$ N. In addition, the critical force leading to buckling is calculated as $F_{cr} = 943.53$ N.

To assess the statics performance of the stage, static structural FEA simulation is carried out by applying an input displacement at the input end. With a 5-mm input displacement, the FEA result is illustrated in Fig. 3.7, which reveals a safety factor of 1.7 for the material. The required actuation force is deduced as 68.83 N. Considering the FEA result as the benchmark value, it is found that the analytical model overestimates the force capability by 25.0%. Additionally, the maximum allowable displacement evaluated by FEA is 8.46 mm. Thus, the analytical model is more conservative because it underestimates the displacement by about 30.9%. It is observed that the simple analytical models predict the stage performance with relatively large deviations. As a future work, nonlinear models may be established to evaluate the stage performances more accurately.

Moreover, modal analysis is conducted to evaluate the dynamics performance of the XY stage. FEA simulation results show that the first three natural frequencies are 48.3, 48.7, and

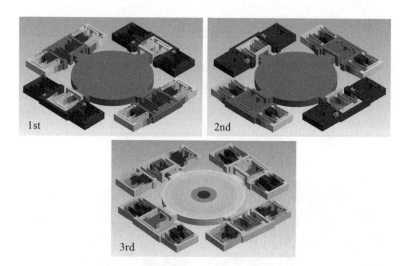

Figure 3.8   FEA simulation results of the first three resonant modes for the XY stage.

100.1 Hz, respectively. As shown in Fig. 3.8, the mode shapes of the first two modes are translations along the two working axes, and the mode shape of the third mode is rotation of the output platform in the working plane. The close values of the first two natural frequencies imply that the XY stage has almost identical dynamics property in the two working axes. The fact that the frequency of the rotational mode is over twice as high as those of the translational modes confirms that the stage has two translational degrees-of-freedom (DOFs).

3.3.2   Prototype Fabrication

The fabricated prototype is shown graphically in Fig. 3.9. The stage output motion in each axis is measured by a laser displacement sensor with a measuring range of 20 mm. The control hardware is constructed by a NI USB-6259 board equipped with A/D and D/A channels.

The FEA simulation results suggest that the actuation force of the VCM should be chosen as

$$68.83 \text{ N} \leq F_{\max}^{\mathrm{VCM}} \leq 754.82 \text{ N} \tag{3.12}$$

so as to ensure a large output motion range (± 5 mm) as well as the safety of the material. In view of the force capability and motion range requirements, the VCM (model: NCC05-18-060-2X, from H2W Techniques, Inc.) is selected to provide a sufficiently large output force of 194.6 N with a stroke of 12.7 mm. In addition, control algorithms are developed with NI LabVIEW software to realize real-time control.

3.3.3   Open-Loop Experimental Results

The motion range and crosstalk of the positioning system in the two working directions are examined through experimental studies. Specifically, a 1-Hz sine wave with 10-V amplitude is used as the input command to conduct motion range testing. When the XY stage is driven by

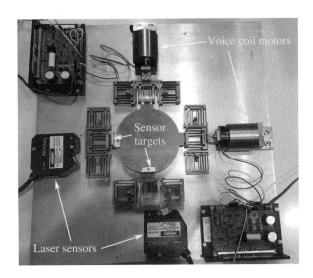

Figure 3.9   The developed prototype of the XY micropositioning stage.

one VCM to translate along the x (or y) direction, the displacements in both axes are measured to determine the motion range and crosstalk value. The experiments have been conducted using open-loop control. The output displacements of the stage are recorded as shown in Fig. 3.10. It is observed that the XY stage has a workspace of about 10.5 mm × 10.5 mm with the maximum crosstalk of 1.6% between the two working axes. Experimental results also exhibit that hysteresis effects exist in the positioning system. The hysteresis mainly arises from the nonlinearity of the employed VCM, which works based on the electromagnetic principle.

Compared with the XY stage previously presented in [2], which has dimensions of 255 mm × 255 mm × 25 mm with a 10-mm motion range, the current design is cable of 10.5 mm × 10.5 mm motion while possessing a much smaller dimension of 214 mm × 214 mm × 10 mm. That is, the developed XY stage has an area ratio of 0.2407%, which is 56.5% larger than the one (0.1538%) reported in [2]. Thus, the presented parallel-kinematic XY stage has a much more compact size. The compactness will be further enhanced by implementing an optimum design in future work.

The experimental results confirm the motion decoupling of the XY stage. In addition, the buckling phenomenon docs not occur during the operation, which validates the effectiveness of the design procedure. It is found that the output decoupling of the XY stage is enabled by employing MCPFs, whereas the secondary stages increase the mass of the moving components, which leads to a not-high natural frequency of the XY stage as indicated by the FEA results.

To suppress the nonlinearity and achieve a precise positioning for the micropositioning system, a closed-loop control scheme is presented in the following section.

3.4   EMPC Control Scheme Design

In this section, an EMPC scheme is developed to realize a precise positioning of the micropositioning system.

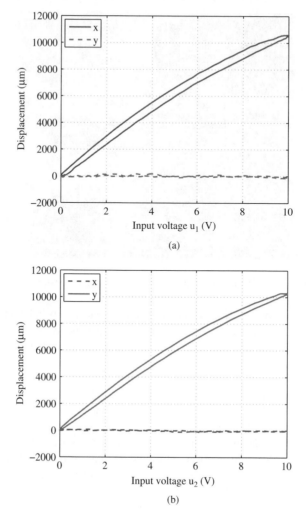

Figure 3.10 Motion range and crosstalk test results with (a) VCM #1 and (b) VCM #2 driven for the XY micropositioning system.

3.4.1 Problem Formulation

The fine decoupling property of the XY stage allows the adoption of a single-input/single-output (SISO) controller for each working axis. The dynamics model of a VCM-driven micropositioning system for one working axis can be described by a linear-time-invariant (LTI) model accompanied by disturbance:

$$\dot{\mathbf{X}}(t) = \mathbf{A}_c\mathbf{X}(t) + \mathbf{B}_c u(t) + \mathbf{f}(t) \tag{3.13}$$

$$y(t) = \mathbf{C}\mathbf{X}(t) + Du(t) \tag{3.14}$$

where the state vector $\mathbf{X} \in \mathbb{R}^n$ $(n \geq 1)$, the output $y \in \mathbb{R}$, the control input $u \in \mathbb{R}$, and the disturbance vector $\mathbf{f} \in \mathbb{R}^n$ is supposed to be smooth and bounded. The system matrices $\mathbf{A}_c$, $\mathbf{B}_c$,

C, and D are constant with approximate dimensions. In addition, the lumped disturbance $\mathbf{f}$ describes the combined effects of unmodeled nonlinearity, higher-order dynamics, and external perturbations.

Using a sampling time T, the continuous-time system model of Eqs. (3.13) and (3.14) can be discretized as follows:

$$\mathbf{X}_{k+1} = \mathbf{A}\mathbf{X}_k + \mathbf{B}u_k + \mathbf{f}_k \tag{3.15}$$

$$y_k = \mathbf{C}\mathbf{X}_k + Du_k \tag{3.16}$$

where $\mathbf{X}_k = \mathbf{X}(kT)$ and

$$\mathbf{A} = e^{\mathbf{A}_c T} \tag{3.17}$$

$$\mathbf{B} = \int_0^T e^{\mathbf{A}_c \tau} d\tau \, \mathbf{B}_c \tag{3.18}$$

$$\mathbf{f}_k = \int_0^T e^{\mathbf{A}_c \tau} \mathbf{f}((k+1)T - \tau) d\tau. \tag{3.19}$$

It is observed that both $\mathbf{B}$ and $\mathbf{f}_k$ are of the order $O(T)$ with respect to the sampling time T. Moreover, the difference equation of the disturbance can be derived as

$$\mathbf{f}_k - \mathbf{f}_{k+1} = \int_0^T e^{\mathbf{A}_c T} \left(\int_{kT-\tau}^{(k+1)T-\tau} \dot{\mathbf{f}}(\theta) d\theta \right) d\tau. \tag{3.20}$$

It is assumed that the triplet $(\mathbf{A}, \mathbf{B}, \mathbf{C})$ of the nominal system is both controllable and observable, with the matrices $\mathbf{B}$ and $\mathbf{C}$ being of full rank. Here, the concerned positioning control problem is how to make the actual output (y_k) track a command reference input (r_k) precisely.

Concerning a flexure-based micropositioning system in the current work, the system identification conducted later reveals that the plant is of NMP. That is, the system has unstable zeros. The transfer function of the system model of Eqs. (3.13) and (3.14) can be expressed as $G(s) = \mathbf{C}(s\,\mathbf{I} - \mathbf{A}_c)^{-1}\mathbf{B}_c + D$. A NMP plant indicates that the inverse of the system model G^{-1} has unstable poles. Thus, some powerful control techniques based on the inversion principle cannot be applied directly. In the sequel, an improved scheme based on the MPC technique is proposed for a precise positioning control of the NMP micropositioning system.

3.4.2 EMPC Scheme Design

In this subsection, an EMPC strategy is presented, which is based on the MPC scheme enhanced with PID control.

3.4.2.1 PID Controller Design

It is well known that PID is a model-free controller, which solves the control command by making use of the control error only. A digital realization of the PID algorithm can be expressed as follows:

$$u_k^{\text{pid}} = K_p e_k + K_i \sum_{i=0}^{k} e_k + K_d(e_k - e_{k-1}) \tag{3.21}$$

with displacement error

$$e_k = r_k - y_k \tag{3.22}$$

where r_k and y_k represent the desired and actual system outputs, and K_p, K_i, and K_d denote the proportional, integral, and derivative gains, respectively.

The key issue of PID control lies in the tuning of the three controller parameters. In this work, the Ziegler–Nichols (Z–N) method is adopted due to its popularity. Based on the digital PID controller, an enhanced MPC strategy is presented below.

3.4.2.2 EMPC Controller Design

By using the conventional MPC strategy, the overall control action can be expressed in a difference form:

$$u_k = u_{k-1} + \Delta u_k. \tag{3.23}$$

In general, the purpose of MPC is to generate the incremental control action Δu_k in an optimal manner. However, the following simulation studies show that the conventional MPC produces a slow transient response with a large steady-state error, which is not suitable for micropositioning applications.

In order to overcome these limitations, an EMPC scheme is proposed by considering u_k^{pid} as the previous-step control effort u_{k-1}. In this way, the controller performance will be significantly improved.

Taking into account Eqs. (3.15) and (3.23) together, a new state-space equation can be obtained:

$$\mathbf{Z}_{k+1} = \bar{\mathbf{A}}\mathbf{Z}_k + \bar{\mathbf{B}}\Delta u_k + \bar{\mathbf{I}}f_k \tag{3.24}$$

$$y_k = \bar{\mathbf{C}}\mathbf{Z}_k \tag{3.25}$$

where

$$\mathbf{Z}_k = \begin{bmatrix} \mathbf{X}_k \\ u_{k-1} \end{bmatrix} \quad \bar{\mathbf{A}} = \begin{bmatrix} \mathbf{A} & \mathbf{B} \\ \mathbf{0} & 1 \end{bmatrix} \quad \bar{\mathbf{B}} = \begin{bmatrix} \mathbf{B} \\ 1 \end{bmatrix} \quad \bar{\mathbf{I}} = \begin{bmatrix} \mathbf{I} \\ 0 \end{bmatrix}$$

$$\bar{\mathbf{C}} = [\mathbf{C} \quad 0]. \tag{3.26}$$

Based on Eqs. (3.24) and (3.25), an N_n-step-ahead prediction of the system output can be derived as follows:

$$\begin{aligned} y_{k+N_n} = {} & \bar{\mathbf{C}}\bar{\mathbf{A}}^{N_n}\mathbf{Z}_k + \bar{\mathbf{C}}\bar{\mathbf{A}}^{N_n-1}\bar{\mathbf{B}}\Delta u_k + \bar{\mathbf{C}}\bar{\mathbf{A}}^{N_n-2}\bar{\mathbf{B}}\Delta u_{k+1} \\ & + \cdots + \bar{\mathbf{C}}\bar{\mathbf{B}}\Delta u_{k+N_n-1} + \bar{\mathbf{C}}\bar{\mathbf{A}}^{N_n-1}\bar{\mathbf{I}}f_k \\ & + \bar{\mathbf{C}}\bar{\mathbf{A}}^{N_n-2}\bar{\mathbf{I}}f_{k+1} + \cdots + \bar{\mathbf{C}}\bar{\mathbf{I}}f_{k+N_n-1} \end{aligned} \tag{3.27}$$

where the positive integer N_n is the prediction horizon.

The N_n prediction equations can be stacked together in the form

$$\mathbf{y}_k^a = \mathbf{\Lambda}\mathbf{Z}_k + \mathbf{\Phi}\Delta\mathbf{u}_{k-1}^a + \mathbf{\Gamma}\mathbf{f}_{k-1}^a \tag{3.28}$$

where the vectors for the future system output, incremental control action, and disturbances are

$$\mathbf{y}_k^a = [y_{k+1},\ y_{k+2},\ \ldots,\ y_{k+N_n}]^T \tag{3.29}$$

$$\Delta\mathbf{u}_{k-1}^a = [\Delta u_k,\ \Delta u_{k+1},\ \ldots,\ \Delta u_{k+N_n-1}]^T \tag{3.30}$$

$$\mathbf{f}_{k-1}^a = [\mathbf{f}_k^T,\ \mathbf{f}_{k+1}^T,\ \ldots,\ \mathbf{f}_{k+N_n-1}^T]^T. \tag{3.31}$$

Additionally, the three matrices take on the following forms:

$$\boldsymbol{\Lambda} = \begin{bmatrix} \bar{\mathbf{C}}\bar{\mathbf{A}} \\ \bar{\mathbf{C}}\bar{\mathbf{A}}^2 \\ \vdots \\ \bar{\mathbf{C}}\bar{\mathbf{A}}^{N_n} \end{bmatrix} \tag{3.32}$$

$$\boldsymbol{\Phi} = \begin{bmatrix} \bar{\mathbf{C}}\bar{\mathbf{B}} & 0 & \cdots & 0 \\ \bar{\mathbf{C}}\bar{\mathbf{A}}\bar{\mathbf{B}} & \bar{\mathbf{C}}\bar{\mathbf{B}} & \cdots & 0 \\ \vdots & \vdots & \ddots & \vdots \\ \bar{\mathbf{C}}\bar{\mathbf{A}}^{(N_n-1)}\bar{\mathbf{B}} & \bar{\mathbf{C}}\bar{\mathbf{A}}^{(N_n-2)}\bar{\mathbf{B}} & \cdots & \bar{\mathbf{C}}\bar{\mathbf{B}} \end{bmatrix} \tag{3.33}$$

$$\boldsymbol{\Gamma} = \begin{bmatrix} \bar{\mathbf{C}}\bar{\mathbf{I}} & 0 & \cdots & 0 \\ \bar{\mathbf{C}}\bar{\mathbf{A}}\bar{\mathbf{I}} & \bar{\mathbf{C}}\bar{\mathbf{I}} & \cdots & 0 \\ \vdots & \vdots & \ddots & \vdots \\ \bar{\mathbf{C}}\bar{\mathbf{A}}^{(N_n-1)}\bar{\mathbf{I}} & \bar{\mathbf{C}}\bar{\mathbf{A}}^{(N_n-2)}\bar{\mathbf{I}} & \cdots & \bar{\mathbf{C}}\bar{\mathbf{I}} \end{bmatrix}. \tag{3.34}$$

Assume that the N_n-step preview ($\mathbf{r}_k^a = [r_{k+1}, r_{k+2}, \ldots, r_{k+N_n}]^T$) of the desired reference input is available. Then, a cost function for minimization can be expressed as

$$J = (\mathbf{r}_k^a - \mathbf{y}_k^a)^T(\mathbf{r}_k^a - \mathbf{y}_k^a) + w(\Delta\mathbf{u}_{k-1}^a)^T\Delta\mathbf{u}_{k-1}^a \tag{3.35}$$

where w is the weighting to limit the partial control action. It is notable that $\Delta\mathbf{u}_{k-1}^a$ approaches zero as $(\mathbf{r}_k^a - \mathbf{y}_k^a)$ tends to zero. Thus, the cost function J vanishes at steady state.

Substituting Eq. (3.28) into the cost function (3.35) and applying the optimization criterion by setting

$$\frac{\partial J}{\partial \Delta\mathbf{u}_{k-1}^a} = 0 \tag{3.36}$$

yields

$$\Delta\mathbf{u}_{k-1}^a = (\boldsymbol{\Phi}^T\boldsymbol{\Phi} + w\mathbf{I})^{-1}\boldsymbol{\Phi}^T(\mathbf{r}_k^a - \boldsymbol{\Lambda}\mathbf{Z}_k - \boldsymbol{\Gamma}\mathbf{f}_{k-1}^a) \tag{3.37}$$

where the future disturbance values in $\mathbf{f}_{k-1}^a$ are unknown. In this work, they are estimated by

$$\hat{\mathbf{f}}_{k-1}^a = [\mathbf{f}_{k-1}^T\ \ \mathbf{f}_{k-1}^T\ \ \cdots\ \ \mathbf{f}_{k-1}^T]^T \tag{3.38}$$

where $\mathbf{f}_{k-1}$ denotes the one-step delayed estimation of the disturbance, which can be obtained by noting Eq. (3.15).

Because only the first element of the predicted control sequence is used, the optimal incremental control action can be generated:

$$\Delta u_k = \mathbf{e}(\boldsymbol{\Phi}^T\boldsymbol{\Phi} + w\mathbf{I})^{-1}\boldsymbol{\Phi}^T(\mathbf{r}_k^a - \boldsymbol{\Lambda}\mathbf{Z}_k - \boldsymbol{\Gamma}\hat{\mathbf{f}}_{k-1}^a) \tag{3.39}$$

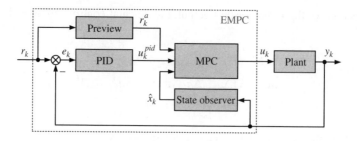

Figure 3.11   Block diagram of the EMPC control scheme.

where the selection vector $\mathbf{e} = [1, 0, 0, \ldots, 0]$.

Therefore, the total control action of the EMPC can be expressed as

$$u_k = u_k^{\text{pid}} + \Delta u_k$$

$$= K_p e_k + K_i \sum_{i=0}^{k} e_k + K_d(e_k - e_{k-1})$$

$$+ \mathbf{e}(\boldsymbol{\Phi}^T \boldsymbol{\Phi} + w\mathbf{I})^{-1} \boldsymbol{\Phi}^T (\mathbf{r}_k^a - \boldsymbol{\Lambda}\mathbf{Z}_k - \boldsymbol{\Gamma}\hat{\mathbf{f}}_{k-1}^a). \tag{3.40}$$

A block diagram of the proposed EMPC controller is shown in Fig. 3.11. It is observed that the N_n-step preview ($\mathbf{r}_k^a$) of the reference input is needed by the EMPC scheme.

3.4.3 State Observer Design

An investigation of the controller (3.40) reveals that its implementation needs the full state ($\mathbf{X}_k$) feedback of the system. However, only partial states are available in practice. Thus, a state observer is designed to estimate the full state by making use of the position information of the micropositioning system provided by displacement sensors.

In the literature, various methods are available to estimate the system state, such as Luenberger observer, high-gain observer, Kalman filter, sliding mode observer, and so on [13]. In this work, a Luenberger observer is adopted owing to its simplicity. The state observer takes on the form

$$\hat{\mathbf{X}}_{k+1} = \mathbf{A}\hat{\mathbf{X}}_k + \mathbf{B}u_k + \mathbf{L}(y_k - \hat{y}_k) + \hat{\mathbf{f}}_k \tag{3.41}$$

$$\hat{y}_k = \mathbf{C}\hat{\mathbf{X}}_k \tag{3.42}$$

where $\mathbf{L}$ is the observer gain vector. The notation $\hat{\mathbf{X}}_k$ represents the estimate of Eq. (3.15) obtained by its one-step delayed estimation:

$$\hat{\mathbf{f}}_k = \mathbf{f}_{k-1} = \hat{\mathbf{X}}_k - \mathbf{A}\hat{\mathbf{X}}_{k-1} - \mathbf{B}u_{k-1}. \tag{3.43}$$

Subtracting Eq. (3.41) from Eq. (3.15) allows the derivation

$$\tilde{\mathbf{X}}_{k+1} = (\mathbf{A} - \mathbf{LC})\tilde{\mathbf{X}}_k + (\mathbf{f}_k - \mathbf{f}_{k-1}) \tag{3.44}$$

where $\widetilde{\mathbf{X}}_k = \mathbf{X}_k - \hat{\mathbf{X}}_k$ denotes the estimation error of the state $\mathbf{X}_k$. It is seen that the state observer is independent of the controller. To guarantee the stability of the observation unit, the observer gain $\mathbf{L}$ should be designed to make all the eigenvalues of the matrix $(\mathbf{A} - \mathbf{LC})$ locate inside the unit circle.

The convergence speed of the observer relies on the location of the poles of the matrix $(\mathbf{A} - \mathbf{LC})$. The poles can be arbitrarily placed by the gain vector $\mathbf{L}$. Generally, a small pole value will give rapid convergence, and the convergence speed of the observer should be faster than the system response, so that the observer dynamics is insignificant compared with the system dynamics.

3.4.4 Tracking Error Analysis

With the proposed controller, the tracking error can be derived as follows:

$$
\begin{aligned}
e_{k+1} &= r_{k+1} - y_{k+1} \\
&= r_{k+1} - \mathbf{CX}_{k+1} \\
&= r_{k+1} - \mathbf{C}[\mathbf{AX}_k + \mathbf{B}(u_k^{\text{pid}} + \Delta u_k) + \mathbf{f}_k] \\
&= r_{k+1} - \mathbf{CAX}_k - \mathbf{CB}u_k^{\text{pid}} - \mathbf{Cf}_k - \mathbf{CB}\Delta u_k.
\end{aligned} \tag{3.45}
$$

Substituting Eq. (3.40) into Eq. (3.45), along with a necessary algebraic operation, results in

$$
e_{k+1} = \mathbf{C}(\mathbf{f}_k - \mathbf{f}_{k+1}) \tag{3.46}
$$

which indicates that the ultimate bound of the output tracking error is bounded by the change rate of the disturbance $\mathbf{f}_k$. In practice, with a sufficiently high sampling frequency, a lower error bound can be achieved under the assumption that the disturbance is bounded and smooth.

3.5 Simulation and Experimental Studies

In this section, the designed controller is verified by a series of control simulation and experimental studies conducted on the micropositioning system.

3.5.1 Plant Model Identification

The micropositioning system is considered as a linear system with bounded disturbances, as described by Eqs. (3.13) and (3.14). The linear plant model is identified by the frequency response method via experiments. Specifically, swept-sine waves with amplitude of 0.05 V and frequency range of 1–300 Hz are applied to the power amplifier to drive one VCM. The position responses of the XY stage in the two working directions are simultaneously recorded using a sampling rate of 2 kHz. With the two VCMs actuated individually, the frequency responses in the two axes are shown in Figs. 3.12 and 3.13, which are obtained by fast Fourier transform

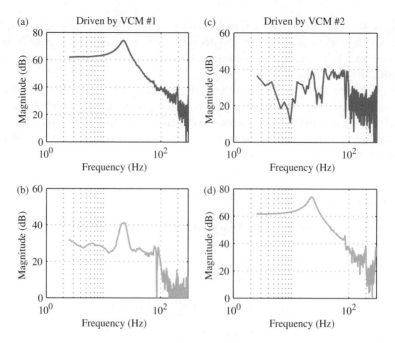

Figure 3.12 Magnitude plots of frequency responses of the XY micropositioning system: (a) G_{x1}, (b) G_{y1}, (c) G_{x2}, and (d) G_{y2}.

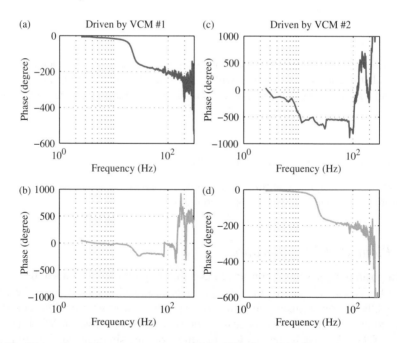

Figure 3.13 Phase plots of frequency responses of the XY micropositioning system: (a) G_{x1}, (b) G_{y1}, (c) G_{x2}, and (d) G_{y2}.

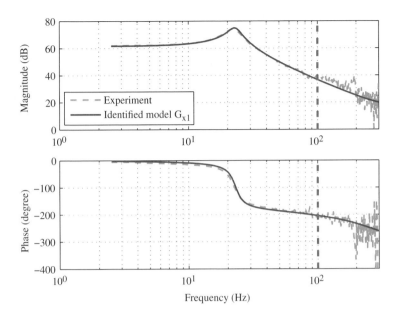

Figure 3.14 Plant frequency responses in the x-axis obtained by experiments and the identified second-order model.

(FFT). The frequency response of the i-axis motion induced by driving VCM k is denoted by $G_{ik}(j\omega) = d_i(j\omega)/u_k(j\omega)$, where the axis index $i = x$ and y, and the actuator index $k = 1$ and 2, respectively. It is observed that with one VCM driven, the response in the passive axis is about 30 dB lower than that in the primary axis in the low-frequency range. This confirms that the two axial motions of the XY stage are well decoupled. Thus, a SISO controller can be designed for each working axis.

To verify the proposed control scheme, only the x-axis motion control is handled in this work. The input–output data sets are used to identify the plant transfer function as shown in Eq. (3.47) by estimating the model from the frequency response data. The frequency responses obtained by experiment and estimated by the identified model (3.47) are compared in Fig. 3.14:

$$G_{x1}(s) = \frac{16.03s^2 + 2.904 \times 10^6 s + 3.122 \times 10^7}{s^2 + 54.01s + 2.504 \times 10^4}. \tag{3.47}$$

It is observed that the first resonant mode occurs around 23 Hz, and the identified second-order model matches the system dynamics well at frequencies below 100 Hz. In order to capture high-frequency dynamics accurately, a model of higher order is demanded. Here, a simple second-order model is employed to demonstrate the effectiveness of the proposed control scheme. The transfer function (3.47) shows that the two zeros (1587.4 $\pm$ 1341.7i) locate on the right-hand side of the s-plane. Thus, the micropositioning system has a NMP plant.

It is observed that the resonant mode (23 Hz) is much lower than the FEA simulation result (48.3 Hz). The discrepancy mainly comes from the mass of the moving coils of the VCM and the sensor targets, which is not considered in the FEA simulation.

3.5.2 Controller Parameter Design

Comparing the state-space model of Eqs. (3.13) and (3.14) with Eq. (3.47) yields the system parameters, which allow the calculation of the discrete-time system model of Eqs. (3.15) and (3.16) with a sampling time T. Generally, to capture the system behavior precisely, a sampling frequency larger than 15 times the first resonant mode (23 Hz) of the system is preferred. In this work, a sampling frequency of 500 Hz is adopted. With $T = 0.002$ s, the discrete-time system is represented by

$$\mathbf{A} = \begin{bmatrix} 0.9005 & 0.0019 \\ -39.7318 & 0.9596 \end{bmatrix}, \quad \mathbf{B} = \begin{bmatrix} 12.8832 \\ 49648.7140 \end{bmatrix}$$

$$\mathbf{C} = [1 \quad 0], \quad D = 5.7656.$$

Based on the generated linear model, the state observer is constructed. The observer gain vector $\mathbf{L}$ is designed by the pole placement technique. The trial simulations suggest that small pole values produce rapid convergence, since a high bandwidth is achieved by the observer. However, experiments reveal that too small a pole leads to instability of the system, which arises from the sensor noises and spill-over effects of the neglected dynamics. To make a tradeoff between the convergence speed and stability, the two poles are assigned as (0.85, 0.85), which are chosen to guarantee a relatively quicker convergence as well as robustness of the observer in the presence of unmodeled dynamics. The corresponding gains are calculated as $\mathbf{L} = [0.1601, 0.0016]^T$. In addition, the PID control gains are tuned by means of the Z–N method, which gives $K_p = 0.0010$, $K_i = 0.0809$, and $K_d = 3.1612 \times 10^{-6}$. Moreover, a prediction horizon $N_n = 10$ is selected for illustration.

3.5.3 Simulation Studies and Discussion

Before conducting experimental investigations, the performance of the designed controller is tested by several simulation studies as carried out below. For demonstration, a 1-mm set-point positioning is carried out to discover the influence of the weight parameter w on the positioning performance.

First, the conventional MPC without PID control is tested. With increasing weight w, the positioning results are depicted in Fig. 3.15(a). It is observed that a quicker response is obtained using a smaller weight, $w = 2 \times 10^7$. However, the control action oscillates as a consequence, which leads to an oscillation in the positioning result. On the contrary, a larger weight $w = 8 \times 10^7$ results in less oscillation in positioning. Nevertheless, this is at the cost of a large steady-state error. Hence, the conventional MPC is not capable of producing a satisfactory positioning result. That is the reason why an EMPC scheme is proposed. For a comparative study, $w = 4 \times 10^7$ is selected to make a tradeoff between the response speed and positioning accuracy. This produces a 5% settling time of 0.059 s with a steady-state root-mean-square error (RMSE) of 5.209 μm.

Second, the performance of the proposed EMPC control scheme is verified. For different weight values, the positioning results are shown in Fig. 3.15(b). As w increases, the control effort Δu_k is more constrained. Thus, it is observed that the EMPC positioning result is more like the stand-alone PID result. The weight value $w = 4 \times 10^5$ gives a better result in terms of

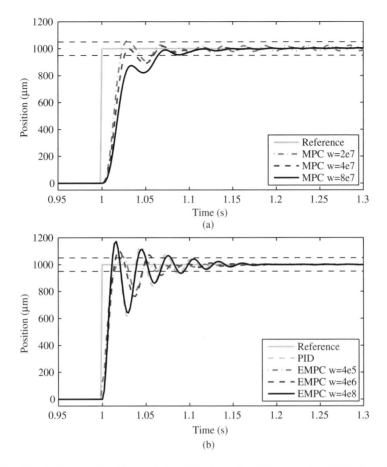

Figure 3.15 Simulation results of set-point positioning using (a) conventional MPC controller and (b) proposed EMPC controller.

a small settling time of 0.046 s and a low steady-state RMSE of 5.574×10^{-5} μm. Compared with the conventional MPC method, the EMPC scheme has reduced the settling time by 22% and the steady-state positioning error by over 10^5 times. The effectiveness of the proposed controller is evident from the positioning results.

In addition, concerning the EMPC with $w = 4 \times 10^5$, the components of the control actions are depicted in Fig. 3.16. It is observed that the incremental control action Δu_k takes effect initially and then vanishes gradually as the time elapses. As a result, the positioning error is gradually reduced to zero. This process is enabled by minimizing the cost function (3.35) using the MPC technique. The results illustrate the efficiency of the EMPC strategy intuitively.

Third, the effect of the prediction horizon selection is investigated. With variation of the prediction horizon N_n, the settling time and steady-state RMSEs are depicted in Fig. 3.17(a) and (b), respectively. It is observed that the larger the prediction horizon, the more rapid the system response. Concerning the RMSE, it appears that $N_n = 4$ leads to the lowest value. For a specific application, a trial-and-error approach is needed to select a suitable prediction horizon.

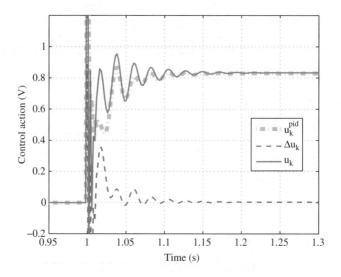

Figure 3.16   Simulation results of control action components for EMPC with $w = 4 \times 10^5$.

The performed control simulation investigations validate the effectiveness of the proposed controller. The control scheme is further verified by experimental studies, as discussed below.

3.5.4   Experimental Results and Discussion

The purpose of the experimental study is to verify whether the control scheme can achieve the effective performance as shown by the simulation study, when the stage works in the actual conditions. So, the emphasis of the experimental study is focused on the transient response speed and steady-state error of the controller.

Specifically, a 1-mm step response of the XY stage is tested for x-axis motion. The result of the experimental study is shown in Fig. 3.18. By comparing the results of the PID and presented EMPC controllers, it is found that both EMPC and PID control lead to a similar variation tendency of the stage's movement. After analyzing the results in more detail, it is observed that the response speed of the proposed EMPC method is a little slower than that of PID control, although the EMPC produces a quicker response than the conventional MPC scheme. Even so, the EMPC scheme exhibits a fast enough transient response speed, which is able to satisfy the requirement of the concerned application. Moreover, the overshoot of the EMPC scheme is smaller than that of PID control.

Then, the steady-state error is examined. The experimental results of the steady-state errors for the PID and EMPC controllers are shown in Fig. 3.19. It is observed that the PID and EMPC schemes produce steady-state RMSEs of 0.581 and 0.289 μm, respectively. In comparison with the PID result, the EMPC approach reduces the positioning error by about 50%. Hence, the EMPC control scheme improves the positioning accuracy significantly, although it produces a slightly lower transient response speed compared with PID control.

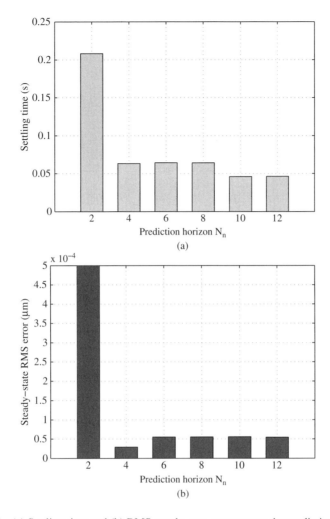

Figure 3.17 (a) Settling time and (b) RMS steady-state error versus the prediction horizon N_n.

It is notable that the controller parameters are not optimally designed. The positioning performance of the designed EMPC controller can be further improved by finely tuning the control parameters through extensive experimental study.

As a model predictive controller, the major restriction of the proposed control scheme lies in the fact that the preview of the reference input is required to implement the control algorithm. This is not feasible when the desired input is unknown beforehand. How to eliminate this limitation deserves future research.

3.6 Conclusion

Based on the concept of multi-stage compound parallelogram flexure, a monolithic parallel-kinematic XY micropositioning stage is designed in this chapter with both a

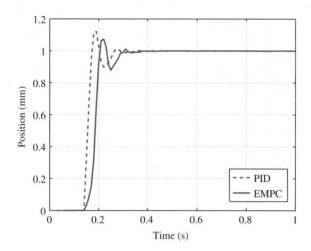

Figure 3.18   Experimental results of set-point positioning obtained by the PID and EMPC controllers.

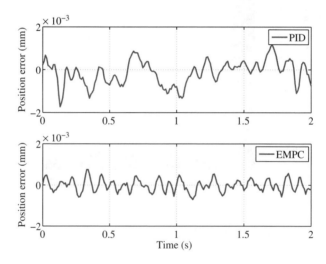

Figure 3.19   Steady-state positioning errors of the PID and EMPC controllers.

compact size and a large workspace. The XY flexure stage produces a workspace of 10.5 mm × 10.5 mm with a compact structure as reflected by the large area ratio of 0.2407%, which is much larger than the existing monolithic parallel-kinematic XY stages. The presented enhanced model predictive control scheme is capable of improving the positioning performance in terms of steady-state error compared with the conventional PID method. Because the realization of the controller does not require the inverse of the plant model, it can easily be extended to both minimum- and non-minimum-phase systems. Although only the leaf springs are used in this work, the concept can be extended to the design of micropositioning systems with any other types of flexure hinges to achieve more types of motion.

References

[1] Kang, D., Kim, K., Kim, D., Shim, J., Gweon, D.G., and Jeong, J. (2009) Optimal design of high precision XY-scanner with nanometer-level resolution and millimeter-level working range. *Mechatronics*, **19** (4), 562–570.

[2] Awtar, S. and Parmar, G. (2010) Design of a large range XY nanopositioning system, in *Proc. ASME 2010 Int. Design Engineering Technical Conf.*, Montreal, Canada, pp. 387–399.

[3] Fatikow, S., Wich, T., Hulsen, H., Sievers, T., and Jahnisch, M. (2007) Microrobot system for automatic nanohandling inside a scanning electron microscope. *IEEE/ASME Trans. Mechatron.*, **12** (3), 244–252.

[4] Hag, J.B. and Bernstein, D.S. (2007) Nonminimum-phase zeros – much to do about nothing – classical control – revisited. Part II. *IEEE Control Syst. Mag.*, **27** (3), 45–57.

[5] Butterworth, J.A., Pao, L.Y., and Abramovitch, D.Y. (2008) The effect of nonminimum-phase zero locations on the performance of feedforward model-inverse control techniques in discrete-time systems, in *Proc. American Control Conf.*, Seattle, WA, pp. 2696–2702.

[6] Ruzbehani, M. (2010) A new tracking controller for discrete-time SISO non minimum phase systems. *Asian J. Control*, **12** (1), 89–95.

[7] Yan, X.G., Edwards, C., and Spurgeon, S.K. (2004) Output feedback sliding mode control for non-minimum phase systems with non-linear disturbances. *Int. J. Control*, **77** (15), 1353–1361.

[8] Deng, M., Inoue, A., Yanou, A., and Hirashima, Y. (2003) Continuous-time anti-windup generalized predictive control of non-minimum phase processes with input constraints, in *Proc. 42nd IEEE Conf. on Decision and Control*, Maui, HI, pp. 4457–4462.

[9] Wills, A.G., Bates, D., Fleming, A.J., Ninness, B., and Moheimani, S.O.R. (2008) Model predictive control applied to constraint handling in active noise and vibration control. *IEEE Trans. Control Syst. Technol.*, **16** (1), 3–12.

[10] Neelakantan, V.A., Washington, G.N., and Bucknor, N.K. (2008) Model predictive control of a two stage actuation system using piezoelectric actuators for controllable industrial and automotive brakes and clutches. *J. Intell. Mater. Syst. Struct.*, **19** (7), 845–857.

[11] Li, Y. and Xu, Q. (2009) Design and analysis of a totally decoupled flexure-based XY parallel micromanipulator. *IEEE Trans. Robot.*, **25** (3), 645–657.

[12] Lobontiu, N. (2002) *Compliant Mechanisms: Design of Flexure Hinges*, CRC Press, Boca Raton, FL.

[13] Koshkouei, A.J. and Zinober, A.S.I. (2002) Sliding mode state observers for discrete-time linear systems. *Int. J. Syst. Sci.*, **33** (9), 751–758.

4

Two-Layer XY Flexure Stage

Abstract: This chapter presents the design and implementation of a modular, decoupled, parallel-kinematic, two-layer flexure XY micropositioning stage with a centimeter range and compact dimension. The XY flexure stage is more compact than existing designs due to its two-layer design. Structural parameters are determined to cater for the requirements on motion range, stiffness, resonant frequency, and payload performances under the influence of the manufacturing tolerance. Finite-element analysis (FEA) simulation reveals a reachable motion range over 20 mm in each working axis. A prototype XY stage is developed for experimental testings. Results exhibit a small crosstalk between the two axes, which indicates good motion decoupling. The implemented proportional-integral-derivative (PID) feedback control enables precision positioning with sub-micrometer resolution and accuracy in an over square-centimeter workspace.

Keywords: Two-dimensional micropositioning, XY flexure mechanisms, Parallel mechanisms, Two-layer design, Modular design, Translational stages, Finite-element analysis, Motion decoupling, Motion control.

4.1 Introduction

The XY micropositioning stage as reported in Chapter 3 is devised as a monolithic structure which is free of assembly. Nevertheless, the monolithic stage occupies a large planar space, in that the space has not been made full use of. The motivation of this chapter is to devise a more compact parallel-kinematic flexure XY stage with a centimeter motion range. Specifically, the concept of a stacked and modular structure is proposed to design a compact flexure stage with parallel-kinematic scheme.

To facilitate the control design for the positioning system, a stage with decoupled output motion is desirable. Additionally, to isolate and protect the actuators, an input decoupling feature is preferable. Hence, the concept of total decoupling [1] is considered in the design procedure. Although the stack concept has been used to design a positioning stage in previous work [2], the stage only delivers a motion range up to 132 μm × 126 μm. Furthermore, it only

Design and Implementation of Large-Range Compliant Micropositioning Systems, First Edition. Qingsong Xu.
© 2016 John Wiley & Sons Singapore Pte Ltd. Published 2016 by John Wiley & Sons Singapore Pte Ltd.

possesses an axial-symmetric structure and has a large crosstalk up to 4.5% between the two working axes. In contrast, a mirror-symmetric structure is proposed in the present work and a negligible parasitic motion is produced. As a consequence, single-input/single-output (SISO) control is easily implemented to achieve precision positioning for the system.

4.2 Mechanism Design

In this section, the mechanism design of a compact precision positioning stage with parallel-kinematic, decoupled and modular structure is presented.

4.2.1 Design of a Two-Layer XY Stage with MCPF

The concept of MCPF is employed to design an XY flexure stage with large workspace. Without loss of generality, $N = 2$ is selected to construct a MCPF, which is then adopted to create a parallel-kinematic XY stage.

To generate a mirror-symmetric structure, the stage is designed as a 4-PP (where P stands for prismatic joint) parallel mechanism as depicted in Fig. 3.4(a), which is evolved from the 2-PP flexure parallel mechanism shown in Fig. 3.2(a). As illustrated in a case study in Chapter 3, this monolithic stage can achieve an area ratio of 0.2407% between the workspace area and the planar dimension area.

To further improve the area ratio so as to get a more compact structure, a basic module structure is presented as shown in Fig. 4.1(a). Once driven by a linear actuator (e.g., VCM) the input displacement of the module is transferred as the pure translation of the cental output platform along the x-axis direction. The output displacement is guided by four MCPFs, which are located at the corners of the module. By mounting the second module on top of the first module, a modular design of the XY stage is generated as depicted in Fig. 4.1(b), where the two stages are arranged in an orthogonal way to construct a 4-PP parallel mechanism. The XY stage features a parallel-kinematic and stacked structure. The modular design allows for cost reduction in terms of manufacturing and maintenance. If the XY stage is driven by VCM #1, a pure translation of the central output stage in the x-axis is produced and guided by the four corner MCPFs of the first module as well as the two other MCPFs of the second module. Similarly, driven by VCM #2, a pure translation of the stage along the y-axis is generated. Hence, a decoupled two-dimensional translation is achieved by the XY stage benefiting from the 4-PP parallel mechanism design.

Moreover, due to a large transverse stiffness of the MCPFs located at the corners, they can tolerate a large transverse load applied in the direction orthogonal to the working axis in the xy-plane, as shown in Fig. 4.1(a). As a result, VCM #1 does not suffer from transverse displacement if the stage is driven by VCM #2. Similarly, VCM #2 does not translate in the transverse direction when the stage is actuated by VCM #1. Therefore, an input decoupling is accomplished by the XY stage. It follows that the proposed XY stage has a total decoupling property.

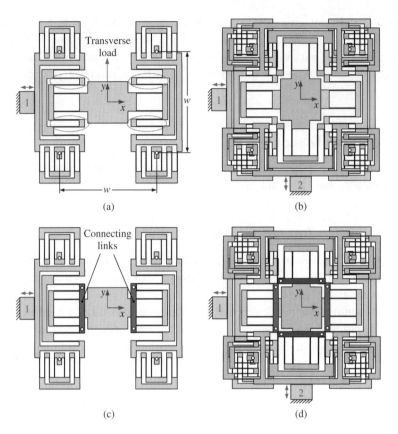

Figure 4.1   (a) A basic module of the modular XY stage, (b) stacked parallel-kinematic XY stage with two basic modules, (c) an enhanced basic module with connecting links, (d) stacked parallel-kinematic XY stage with two enhanced basic modules.

As a result, the stage possesses a much more compact size as shown in Fig. 4.1(b), which indicates a larger area ratio than the previous one depicted in Fig. 3.4(b).

4.2.2   Structure Improvement of the XY Stage

Once the XY stage is driven by a linear actuator (e.g., VCM #1) the leaf flexures in the two centered MCPFs, which are denoted by dashed ellipses in Fig. 4.1(a), suffer from bending/buckling deformations. The undesirable bent flexures cause reduced translational displacement of the stage output platform.

In order to overcome this shortcoming, two out-of-plane connecting links are used to joint the secondary stages of each basic module together as shown in Fig. 4.1(c), which is called an enhanced basic module. The secondary stages can be jointed together because they undergo identical displacements. As a result, an improved structure of the XY stage is obtained as illustrated in Fig. 4.1(d). Because the bending/buckling deformations of the concerned flexures are eliminated, the improved design is able to deliver a much larger output displacement than the original one shown in Fig. 4.1(b).

Due to the mirror-symmetric structure design, the in-plane parasitic motion of the XY stage can be neglected.

4.3 Parametric Design

To develop an XY stage with desired characteristics in terms of motion range, stiffness, load capability, and resonant frequency, the key structural parameters (l, h, and b) are carefully designed as shown in Table 4.1. In addition, for the sake of easy assembly, the four mounting holes of the basic module are designed to form a square with edge length w, as shown in Fig. 4.1(a).

Taking into account that the XY stage has a well-decoupled structure, its main performance can be evaluated by examining the translation along a single working axis. In the parametric design carried out below, it is assumed that the axial deformations of the leaf flexures are neglected and only the bending deformations are considered.

4.3.1 Motion Range Design

The maximum one-sided translation of the XY stage in each axis can be determined by substituting $N = 2$ into Eq. (2.12):

$$d_{max} = \frac{2\sigma_y l^2}{3Eh}. \tag{4.1}$$

It follows that the motion range in each working axis is $\pm d_{max}$, which produces a reachable workspace of $2d_{max} \times 2d_{max}$ with the boundaries defined by the yield stress σ_y. In view of the main parameters as shown in Tables 4.1 and 3.2, $d_{max} = 5.85$ mm is calculated.

To guarantee the safety of the material, the actual workspace should locate inside the reachable workspace as calculated above. That is, for an assigned maximum one-sided translation $D_{max} = 5$ mm of the stage, the parameters should be designed to meet the condition

$$\frac{2\sigma_y l^2}{3Eh} \geq D_{max}. \tag{4.2}$$

The FEA simulation conducted with ANSYS predicts that $d_{max}^{FEA} = 11.08$ mm, which reveals a usable workspace over 20 mm × 20 mm. Moreover, with an assigned $D_{max} = 5$ mm, the

Table 4.1 Main structural parameters of the two-layer XY parallel stage

Parameter	Symbol	Value	Unit
Length of the MCPF flexure	l	25.0	mm
In-plane width of the MCPF flexure	h	0.5	mm
Out-of-plane thickness of the MCPF flexure	b	10.0	mm
Distance between mounting holes	w	77.0	mm

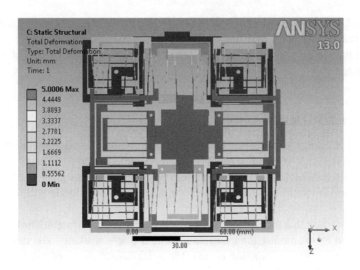

Figure 4.2   Static FEA simulation result of the deformation induced by an input displacement of 5 mm.

deformation generated by static structural analysis is shown in Fig. 4.2. It is found that the maximum stress caused is 226.9 MPa. Hence, the safety factor of the material is calculated as 503 MPa/226.9 MPa = 2.2.

4.3.2   Stiffness and Actuation Force Design

The stiffness of the MCPF in its working axis is determined by substituting $N = 2$ into Eq. (2.11):

$$K_2 = \frac{Ebh^3}{2l^3}.$$

(4.3)

Once driven by a linear actuator (e.g., VCM #1) the deformation of the XY stage is experienced by the six MCPFs which connect the stage output platform to the fixed base in parallel. Therefore, the stage stiffness in each working direction is derived as

$$K = 6K_2 = \frac{3Ebh^3}{l^3}.$$

(4.4)

In order to generate a desired motion range by using VCMs, the stage should be sufficiently compliant so that the elastic deformation energy can be overcome by the driving force of the VCMs. Given the required motion range of $\pm D_{\max}$, the desired maximum driving force can be obtained as

$$F_{\max} = KD_{\max} = \frac{3Ebh^3 D_{\max}}{l^3}.$$

(4.5)

To ensure that the stage is compliant enough, the parameters should be designed to satisfy the condition

$$\frac{3Ebh^3 d_{\max}}{l^3} \leq F_{\text{VCM}}.$$

(4.6)

By selecting the VCM with maximum output force of 194.6 N, the designed structural parameters as shown in Table 4.1 result in the relationship $86.0\,\text{N} \le F_{VCM}$. Hence, the stiffness requirement is met by the parametric design.

In addition, the FEA result indicates that $F_{max}^{FEA} = 102.7\,\text{N} < F_{VCM}$, which also confirms the feasibility of the parametric design.

4.3.3 Critical Load of Buckling

Referring to Fig. 4.2, it is observed that the leaf flexures associated with the MCPFs, which have extra connecting links, suffer from axial loading during operation. Due to a slender architecture, the elastic buckling may occur under compressive loads. Buckling causes instability of the structure and leads to a decreased motion range of the stage. Hence, the maximum force F_{max}, which produces the desired motion range, should not induce buckling deformation of the leaf flexures. The critical load is calculated as follows.

The critical compressive load, which induces elastic buckling of the leaf flexure, can be calculated as

$$P = \frac{\pi^2 EI}{l_{cr}^2} \qquad (4.7)$$

where $I = bh^3/12$ is the area moment of inertia of the cross-section. In addition, the critical length l_{cr} is determined

$$l_{cr} = kl \qquad (4.8)$$

where the coefficient k takes a value from 0.5 to 2 depending on the boundary conditions [3].

Because the two secondary ends of each MCPF are jointed by a connecting link as shown in Fig. 4.1(c), the concerned flexures can be considered to have fixed–fixed constraints. For a beam with fixed–fixed boundary condition, the coefficient k is taken as 0.5. Hence, the critical load can be computed as follows:

$$F_{cr} = 2P = \frac{8\pi^2 EI}{l^2}. \qquad (4.9)$$

In order not to excite the elastic buckling deformation during movement of the stage within the motion range, the following condition should be met:

$$F_{VCM} \le F_{cr}. \qquad (4.10)$$

That is,

$$F_{VCM} \le \frac{2\pi^2 Ebh^3}{3l^2}. \qquad (4.11)$$

In this work, the relationship is calculated as $F_{VCM} \le 943.5\,\text{N}$. Obviously, the maximum actuation force of 194.6 N will not induce elastic buckling deformation of the flexures.

4.3.4 Resonant Frequency

To generate a high bandwidth of the servo system, a high natural frequency of the stage is desired. Based on Lagrange's equation, the free motion of the XY stage can be described by the dynamics equation:

$$\mathbf{M}\ddot{\mathbf{q}} + \mathbf{K}\mathbf{q} = \mathbf{0} \tag{4.12}$$

where $\mathbf{M} = \text{diag}\{M, M\}$ and $\mathbf{K} = \text{diag}\{K, K\}$ are the matrices of the equivalent mass M and stiffness K, respectively. In addition, $\mathbf{q} = [x \quad y]^T$ denotes a vector of the generalized coordinates x and y (i.e., the displacements in the two working axes).

Based on the theory of vibrations, the mode equation can be derived as

$$(\mathbf{K} - \omega_i^2 \mathbf{M})\Phi_i = \mathbf{0} \tag{4.13}$$

where the eigenvector Φ_i, with $i = 1$ and 2, represents a mode shape and the eigenvalue ω_i^2 describes the corresponding natural cyclic frequency, which can be obtained by solving the characteristic equation

$$|\mathbf{K} - \omega_i^2 \mathbf{M}| = \mathbf{0}. \tag{4.14}$$

Then, the natural frequency is computed as $f_i = \frac{1}{2\pi}\omega_i$ in units of hertz. Specifically, $f_1 = f_2 = 54.6$ Hz is calculated for the XY stage.

In addition, the first six resonant frequencies as predicted by modal analysis with ANSYS are shown in Table 4.2, and the corresponding mode shapes are illustrated in Fig. 4.3. The FEA results reveal that the shapes of the first two modes are contributed by the translations along the two working axes, respectively. The third mode is an in-plane rotation with a resonant frequency of 128.5 Hz, which is over twice as high as the first two frequencies. This indicates that a robust translational motion along the working directions is produced.

4.3.5   Out-of-Plane Payload Capability

For practical applications, some loads will be carried by the output platform of the XY stage. The out-of-plane payload that can be supported by the output platform is assessed by carrying out FEA simulations. With a load of 20 kg applied on the output platform, the induced out-of-plane displacement is only 0.4 mm. At the same time, the maximum stress experienced by the material arrives at 82.7 MPa, which is far less than the yield strength of the material. Hence, the XY stage exhibits a good out-of-plane payload capability.

The FEA simulation result of static structural analysis reveals that the stiffness of the XY stage in each working direction is 2.05×10^4 N/m in the absence of out-of-plane load. With an

Table 4.2   First six resonant frequencies of the two-layer XY stage

Mode sequence	Analytical result (Hz)	FEA result (Hz)
1	54.6	59.9
2	54.6	60.5
3	—	128.5
4	—	215.6
5	—	217.7
6	—	219.2

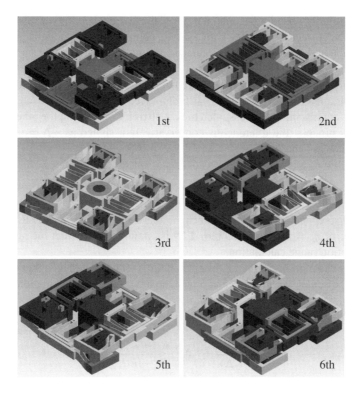

1st 2nd

3rd 4th

5th 6th

Figure 4.3 First six resonant mode shapes of the two-layer XY stage.

out-of-plane load of 20 kg applied on the output platform, the stiffness is assessed as 2.34×10^4 N/m in the working axes. Hence, compared with the free-of-load stiffness, a 20-kg load causes an increase of 14% in the actuation stiffness. In order to produce a motion range of 10 mm under such an external load, the maximum actuation force of 117.2 N is required, which is smaller than the force capability of 194.6 N for the VCMs. Therefore, an out-of-plane load of 20 kg can easily be sustained without influencing the achievement of centimeter range for the XY stage.

4.3.6 Influences of Manufacturing Tolerance

The preceding analyses reveal that the parametric design leads to an XY stage with the desired characteristics. However, the actual values of the parameters are determined by the manufacturing tolerance. Hence, it is important to ensure that the nominal parameters with the uncertainty of tolerances also produce the desired characteristics for the XY stage.

Compared with FEA simulations, the established analytical models are computationally more effective. However, there are certain discrepancies between the analytical results and FEA outputs. The discrepancies are mainly induced by the assumption adopted in the modeling procedure, which only considers the bending deformations of the flexures. In contrast, more accurate results are produced by FEA simulation because all of the possible deformations, including bending and axial deformations, are taken into account. The analytical models can be corrected by multiplying specific compensation factors. Hence, the derived models are employed to discover the influences of manufacturing tolerance on the stage performances.

For a given plate material, its thickness b is a fixed value (e.g., 10 mm). Hence, no tolerance is assigned to parameter b. The tolerances of the employed wire-electrical discharge machining (EDM) and drilling machining are assumed to be ± 10 μm. Considering the other errors caused in the assembly process, the tolerance is conservatively assigned as $\pm \Delta = \pm 30$ μm for the structural parameters l and h.

4.3.6.1 Motion Range

Considering a discrepancy of -47.2% between the analytical prediction and FEA result, a compensation factor of $\eta_d = 1/(1 - 47.20\%) = 1.89$ is employed to correct the analytical model (4.1) for the motion range calculation.

By examining the motion range d_{max} as given in Eq. (4.1), it is deduced that the lower value is produced if the actual parameters l and h take the lower and upper deviations, respectively. On the contrary, the upper value is achieved if l and h are machined with the upper and lower deviations, respectively. That is,

$$\frac{2\sigma_y(l - \Delta)^2 \eta_d}{3E(h + \Delta)} \le d_{max}^c \le \frac{2\sigma_y(l + \Delta)^2 \eta_d}{3E(h - \Delta)} \tag{4.15}$$

where $d_{max}^c = \eta_d d_{max}$ denotes the analytical result corrected by the compensation factor η_d.

In this work, the boundary values are calculated as

$$10.40 \text{ mm} \le d_{max}^c \le 11.78 \text{ mm} \tag{4.16}$$

which means that the lower bound of the motion range still satisfies the design requirement of $d_{max} > D_{max} = 5$ mm.

4.3.6.2 Actuation Force

Regarding the model of the maximum driving force F_{max} as shown in Eq. (4.5), it is compensated by a factor of $\eta_F = 1/(1 - 16.26\%) = 1.19$ to produce a result as accurate as the FEA simulation outcome.

Taking into account Eq. (4.5), the range of the maximum actuation force can be derived as

$$\frac{3Eb(h - \Delta)^3 D_{max} \eta_F}{(l + \Delta)^3} \le F_{max}^c \le \frac{3Eb(h + \Delta)^3 D_{max} \eta_F}{(l - \Delta)^3} \tag{4.17}$$

i.e.,

$$84.7 \text{ N} \le F_{max}^c \le 122.4 \text{ N} \tag{4.18}$$

where $F_{max}^c = \eta_F F_{max}$ represents the compensated analytical result.

It is seen that the upper bound of the maximum actuation force still meets the requirement of $F_{max} < F_{VCM} = 194.6$ N.

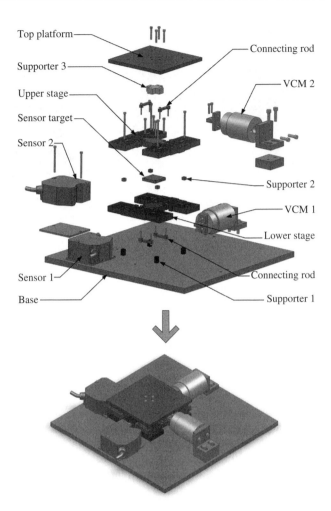

Figure 4.4 Exploded and assembled views of the CAD model for the two-layer XY micropositioning stage.

The foregoing analysis demonstrates that the large motion range of the XY stage is still well guaranteed and the stage can be sufficiently driven by the selected VCMs under the influence of the manufacturing tolerance. Therefore, the parametric design satisfies the requirements in the present work. A CAD model of the developed XY stage is illustrated in Fig. 4.4. The stage is driven by two VCMs, and the output position is measured by two laser displacement sensors.

4.4   Experimental Studies and Results

In this section, a prototype XY precision positioning stage is developed and its performance tested by carrying out a series of experimental studies.

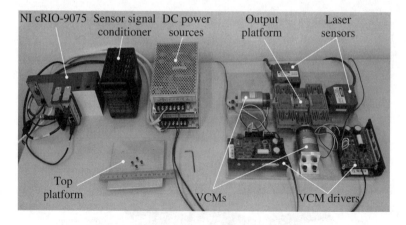

Figure 4.5   Experimental setup of the two-layer XY micropositioning system.

4.4.1   Prototype Development

The experimental setup of the developed XY precision positioning system is shown in Fig. 4.5. Because the two basic modules are designed with identical parameters, they are fabricated using two pieces of Al 7075 plate at one time by the wire-EDM approach. This leads to a reduction in manufacturing cost, which is one of the advantages of the modular design. The XY stage possesses dimensions of 120 mm × 120 mm × 25 mm. The stage is driven by two VCMs. Each VCM delivers a stoke of 12.7 mm with maximum driving force 194.6 N. The position output is measured by two laser displacement sensors (model: LK-H055, from Keyence Corp.), which provide a measurement range of 20 mm. In order to eliminate Abbe errors, the sensors are mounted such that the two measuring directions are coincident with the two working axes of the stage, respectively. In addition, an NI cRIO-9075 real-time controller (from National Instruments Corp.) equipped with NI-9263 analog output module and NI-9870 RS232 serial interface module is adopted to produce excitation signals and acquire the digital sensor readings, respectively. The NI cRIO-9075 combines a real-time processor and a reconfigurable field-programmable gate array (FPGA) within the same chassis, which is connected to a computer via the Ethernet port for communication. Moreover, LabVIEW software is employed to implement real-time control of the micropositioning system.

4.4.2   Statics Performance Testing

By applying a 0.2-Hz low-frequency sinusoidal voltage signal to each VCM, the stage displacements in the two working axes are measured using the two laser sensors. Specifically, by driving VCM #1 with a conservative voltage amplitude of ±6 V, a motion range of 11.75 mm is achieved for the x-axis positioning as shown in Fig. 4.6(a), which is larger than the desired value of $2D_{max} = 10$ mm. In addition, the induced parasitic motion in the y-axis is depicted in Fig. 4.6(b). This shows that the parasitic motion range (146.14 μm) is about 1.24% of the primary motion range along the x-axis.

Similarly, by applying the same input signal to VCM #2, the testing result of the motion range in the y-axis is shown in Fig. 4.6(d). It is observed that the y-axis range is 11.66 mm,

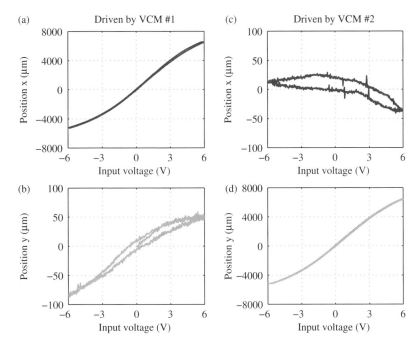

Figure 4.6 Motion range testing results with a 0.2-Hz sinusoidal signal input: (a) and (b) results obtained with VCM #1 driven; (c) and (d) results obtained with VCM #2 driven.

which is also greater than the desired value (10 mm). In addition, the induced parasitic motion in the x-axis is shown in Fig. 4.6(c), which reveals that the maximum crosstalk is 72.17 μm (i.e., 0.62% of the primary motion in the y-axis).

The area ratio of the workspace to planar dimension of the XY stage is calculated as 0.9514%, which is almost four times higher than the result of 0.2407% achieved in previous work [4]. In other words, the developed two-layer XY stage is about four times as compact as the previous one. Evidently, the design objective of a compact XY stage with decoupled motion of centimeter range is well achieved.

Additionally, under open-loop control, the position–voltage relations of the primary motion are nonlinear, as shown in Fig. 4.6(a) and (d). Certain hysteresis effects are present in the parasitic motion, as exhibited in Fig. 4.6(b) and (c). The presence of nonlinearities necessitates the implementation of control techniques in order to achieve precision positioning.

4.4.3 Dynamics Performance Testing

The open-loop dynamics performance of the XY positioning system is tested by the frequency response method. Specifically, swept-sine waves with amplitude of 0.1 V and frequency range of 1–200 Hz are produced by D/A channels of the NI-9263 module to drive the VCMs. Spectral analysis is then conducted to obtain the frequency responses of the stage in the two working axes.

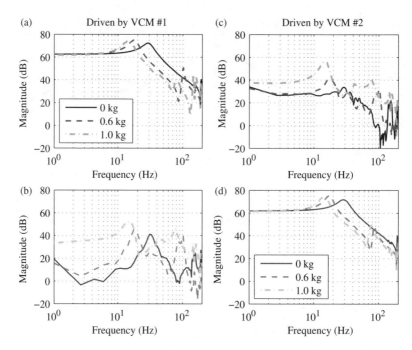

Figure 4.7 Magnitude plots of frequency responses of the XY precision positioning system: (a) G_{x1}, (b) G_{y1}, (c) G_{x2}, and (d) G_{y2}. G_{ni} means the response in the n-axis excited by the ith motor.

By applying the swept-sine signal to VCM #1, the magnitude plots of frequency responses associated with x- and y-axis motion are illustrated in Fig. 4.7(a) and (b) (solid lines), respectively. The corresponding phase plots of the frequency responses are shown in Fig. 4.8(a) and (b) (solid lines), respectively. It is observed that the most significant coupling of the two axes occurs around the resonant frequency. At lower frequencies below the resonant frequency, the coupling effect appears weaker and can be characterized by static crosstalk. Hence, in this work, the dynamic crosstalk is defined as the difference between the magnitudes of frequency responses at the resonant frequency for parasitic and dominant motion. It is observed that the response in the y-axis is about -33 dB lower than that in the primary x-axis at the resonant frequency.

Similarly, driven by VCM #2 alone, the magnitude plots of frequency responses of the XY stage in the two working axes are plotted in Fig. 4.7(c) and (d), respectively. The corresponding phase plots of the frequency responses are shown in Fig. 4.8(c) and (d), respectively. It is seen that the x-axis response is -37 dB lower than the primary y-axis response at the resonant frequency. Hence, a small magnitude of dynamic crosstalk in both working axes of the XY stage is demonstrated. It is notable that in the aforementioned open-loop testings, when VCM #1 (or #2) is driven, the other VCM is set free and no feedback control is applied.

From the results as shown in Fig. 4.7(a) and (d), it is observed that the resonant frequencies for the x- and y-axis motion are 29.3 and 29.6 Hz, respectively. It is found that the frequencies obtained by experiment are lower than the simulation results, as described in Table 4.2. The discrepancy mainly comes from the mass of moving coils of the VCMs, which is not

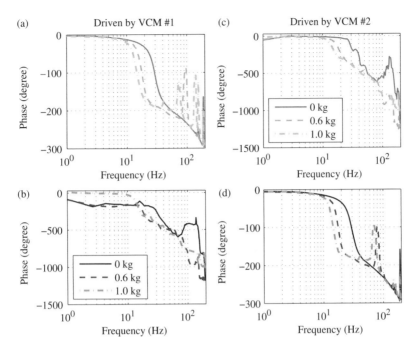

Figure 4.8 Phase plots of frequency responses of the XY precision positioning system: (a) G_{x1}, (b) G_{y1}, (c) G_{x2}, and (d) G_{y2}. G_{ni} means the response in the n-axis excited by the ith motor.

considered in FEA simulations. Besides, the similar resonant frequency of the x- and y-axes indicates almost identical dynamic properties in the two working directions.

Moreover, to discover the influence of the load effect on the stage performance, the frequency responses of the XY stage are experimentally tested under different out-of-plane loads. The magnitude responses for three load cases of 0, 0.6, and 1.0 kg are illustrated in Fig. 4.7. Figure 4.7(a) and (d) shows that under different loads, the magnitude responses in the x- and y-axes at lower frequencies (e.g., 1 Hz) are not altered. This indicates that the motion ranges along the two working axes at low frequencies (e.g., less than 1 Hz) are not affected by the loads.

As the load varies from 0, 0.25, 0.6. 0.8, to 1.0 kg, the dynamic crosstalk and resonant frequencies of the two axes are shown in Fig. 4.9(a) and (b), respectively. It is found that as the external load increases, the resonant frequency is reduced gradually and the dynamic coupling between the two axes becomes worse at the resonant frequencies. In order to alleviate the dynamic coupling effect, a suitable controller is needed to mitigate the interference between the two working axes.

4.4.4 Positioning Performance Testing

Due to the decoupled open-loop static and dynamic responses of the XY stage as verified before, the micropositioning system can be controlled by resorting to two SISO controllers. For illustration, the PID control is adopted to achieve precision positioning owing to its popularity.

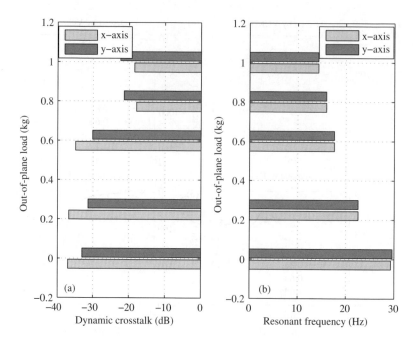

Figure 4.9   (a) Dynamic crosstalk and (b) resonant frequency versus the out-of-plane load.

Two PID controllers are implemented for the two working axes, respectively. The control gains are tuned to produce no overshoot by the Ziegler–Nichols (Z–N) method through experimental studies.

Using the designed controllers, the experimental results of simultaneous set-point positioning of the two working axes are generated as shown in Fig. 4.10. It is observed that the 5% settling time for the x- and y-axes is 0.18 and 0.14 s, respectively, obtained without overshoot effect. In addition, it is seen that the maximum magnitude of response in the perpendicular axis accounts for 0.9% of the response in the dominant axis. Hence, the positioning in one working axis poses a small influence on the other axis, which reveals a robust closed-loop positioning in the two working axes.

Next, a 200-nm consecutive step positioning is tested for each axis and the results are illustrated in Fig. 4.11. The fact that the step size can be clearly identified reveals that the positioning resolution is better than 200 nm for each working axis. Moreover, Fig. 4.12 shows histograms of the positioning errors in the two axes. By calculating the standard deviation (σ) of the errors in the two axes, the 3σ confidence intervals, which correspond to confidence levels of 99.7%, are also depicted in Fig. 4.12. Hence, we can be 99.7% confident that the positioning error will fall within the range between -3σ and 3σ. Experimental results show that the 3σ positioning accuracy is 0.283 μm and 0.254 μm for the x- and y-axes, respectively.

4.4.5   Contouring Performance Testing

To demonstrate the cooperative positioning performance of the two axes for the XY stage, the contouring capability of the stage is tested. For illustration, two circular contouring results are

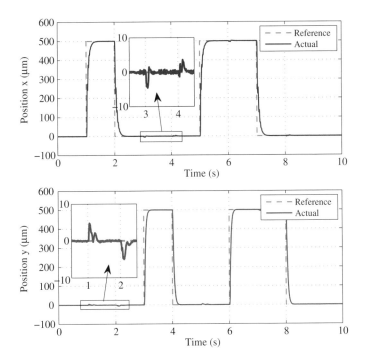

Figure 4.10 Simultaneous set-point positioning results of the two axes for the XY stage.

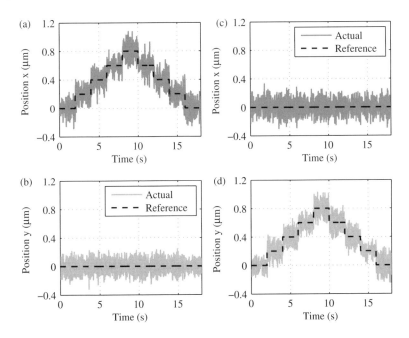

Figure 4.11 Resolution test results of (a), (b) *x*- and (c), (d) *y*-axes, which reveal a positioning resolution of 200 nm in each axis.

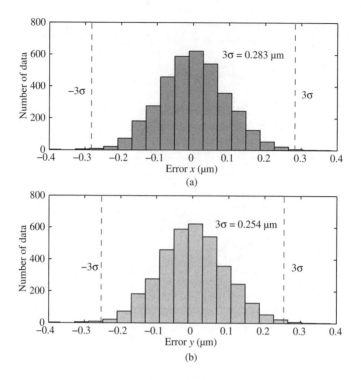

Figure 4.12   Histograms of set-point positioning errors in (a) x-axis and (b) y-axis.

shown in Fig. 4.13. For a circle of 50 μm radius as shown in Fig. 4.13(a), the time history of positioning errors is depicted in Fig. 4.13(b). It is observed that the maximum positioning error accounts for 0.96% of the motion range of the two axes.

In addition, the circular contouring results with a radius of 200 μm are plotted in Fig. 4.13(c) and (d). It is observed that the maximum positioning error is equivalent to 0.83% of the motion range. The contouring accuracy can be further improved by implementing advanced control schemes.

4.4.6   Control Bandwidth Testing

The control bandwidth of the XY stage is tested via the frequency response method. Specifically, by applying a sinusoidal signal with amplitude of 10 μm along with a varying frequency (0.1–30 Hz), the closed-loop frequency responses in the x- and y-axes are shown in Fig. 4.14(a), (b) and (c), (d), respectively. It is observed that there are large phase lags over 90° within the ordinary −3 dB bandwidth, which lead to large tracking errors. Hence, the closed-loop control bandwidth is defined as the frequency at which the phase lag arrives at 30° in this work. With the PID control, a 30°-lag bandwidth of 5.9 and 5.1 Hz is achieved for the x- and y-axes, equivalent to 20% and 17% of the resonant frequencies, respectively. At these cutoff frequencies, the magnitude responses in the x- and y-axes exhibit small values of 0.03 dB and 0.56 dB, respectively.

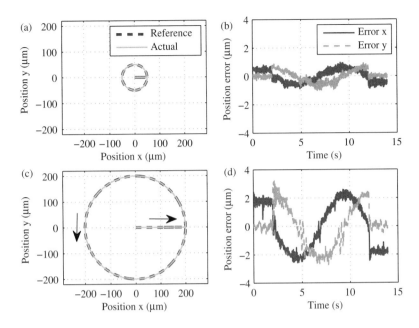

Figure 4.13 Circular contouring results with the radii of (a), (b) 50 µm and (c), (d) 200 µm.

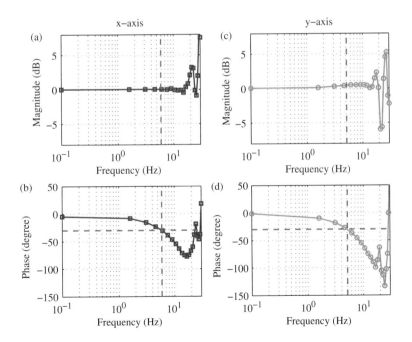

Figure 4.14 Frequency responses of the control system for (a), (b) x-axis and (c), (d) y-axis motion.

Table 4.3 Main performances of the XY precision positioning system without load

Parameter	x-axis	y-axis
Dimension (mm)	120	120
Motion range (mm)	11.75	11.66
Static crosstalk (%)	1.24	0.62
Dynamic crosstalk (dB)	−37	−33
Resonant frequency (Hz)	29.3	29.6
30°-lag control bandwidth (Hz)	5.9	5.1
Resolution (nm)	200	200
3σ accuracy (nm)	283	254

4.4.7   Discussion and Future Work

The main performances of the XY positioning system are summarized in Table 4.3. It is notable that although the actual area ratio is 0.9514% due to the limited strokes of the employed VCMs, an area ratio of 2.7778% is achieved by considering the reachable workspace of 20 mm × 20 mm. In addition, the stage architectural parameters are not optimized. In the future, a more compact stage will be produced by resorting to an optimum parametric design.

To improve the out-of-plane stiffness of the XY flexure stage, an out-of-plane mechanism can be added to connect the output platform on the base [5, 6]. Recently, a compact XY stage is devised by combining an out-of-plane mechanism and an in-plane mechanism together [7]. Although the presented XY stage exhibits a larger area ratio, the actuators for producing the two axial motions are not involved yet. It is noted that the area ratio is calculated without considering the size of the actuators. Generally, VCM provides a larger stroke than PSA at the cost of larger physical size. However, to achieve the same long stroke (e.g., 10 mm), the length of PSA (typically 10 m) is much bigger than that of VCM (fully retracted length of 58 mm in this work). From this point of view, VCM is more suitable for large-range positioning applications.

Although a two-layer structure is employed in the XY stage, both layers are fixed at the base through the four mounting holes as shown in Fig. 4.1. That is, both layers carry an identical mass of moving components. Hence, the stage belongs to a parallel-kinematic architecture, which results in similar frequencies in the two working axes. The resonant frequency of about 30 Hz is relatively low for high-speed positioning applications. Thus, the large motion ranges are achieved at the cost of low resonant frequency. The resonant frequency can be improved by conducting an optimum parametric design of the stage or employing VCMs with a smaller mass of moving coils.

During the operation of the positioning stage, it is important to prevent the occurrence of plastic deformation to ensure the safety of the material. The conducted FEA simulation reveals that the actual motion range, which is produced by the employed VCMs, leads to a safety factor of about 1.9 for each working axis. When the output platform translates within the workspace of the stage, the safety of the material is guaranteed by the safety factor greater than 1. Otherwise, mechanical stoppers may be added to restrict the motion range in each working direction to avoid plastic deformation of the flexures.

In addition, the positioning results of the foregoing experiments are obtained by employing low-pass filters with a cutoff frequency of 30 Hz to reduce the noise of the laser sensors. Actually, the positioning resolution and accuracy are dependent on the performance of the employed displacement sensors. In the future, displacement sensors with higher resolution and lower noise will be adopted to further improve the positioning performance for the XY precision positioning system. Besides, advanced control techniques [8, 9] will be implemented to achieve better accuracy and higher bandwidth for the precision positioning system.

4.5   Conclusion

The design, fabrication, and testing of a large-range XY precision positioning system have been presented in this chapter FEA simulation results show that a reachable workspace of 20 mm × 20 mm is obtained, which indicates an area ratio (workspace size to planar dimension of the stage) of 2.7778%. Due to hardware constraints, a workspace range of 11.75 mm × 11.66 mm is generated. This corresponds to an area ratio of 0.9514%, which is still better than in previous work. In addition, the buckling effect is eliminated, and the stage has an out-of-plane payload capability over 20 kg. Experimental studies on the prototype system reveal a static crosstalk less than 1.3% and a dynamic coupling lower than −33 dB, which indicates a decoupled motion between the two working axes. The dynamics characteristics in both axes are almost identical, as revealed by the similar resonant frequencies around 30 Hz. By realizing PID control for each working axis, a resolution of 200 nm and positioning accuracy better than 340 nm are accomplished. The experimental results confirm the effectiveness of the developed micropositioning system as well as its promising microrobotic applications, with centimeter range and sub-micrometer accuracy.

References

[1] Li, Y. and Xu, Q. (2009) Design and analysis of a totally decoupled flexure-based XY parallel micromanipulator. *IEEE Trans. Robot.*, **25** (3), 645–657.
[2] Li, Y. and Xu, Q. (2011) A novel piezoactuated XY stage with parallel, decoupled, and stacked flexure structure for micro-/nanopositioning. *IEEE Trans. Ind. Electron.*, **58** (8), 3601–3615.
[3] Lobontiu, N. (2002) *Compliant Mechanisms: Design of Flexure Hinges*, CRC Press, Boca Raton, FL.
[4] Xu, Q. (2012) New flexure parallel-kinematic micropositioning system with large workspace. *IEEE Trans. Robot.*, **28** (2), 478–491.
[5] Hao, G. and Kong, X. A novel large-range XY compliant parallel manipulator with enhanced out-of-plane stiffness. *J. Mech. Design*, **134**, 061 009.
[6] Shang, J., Tian, Y., Li, Z., Wang, F., and Cai, K. A novel voice coil motor-driven compliant micropositioning stage based on flexure mechanism. *Rev. Sci. Instrum.*, **86** (9), 095–001.
[7] Huang, C.M. and Su, H.J. (2015) Design of a compliant XY positioning stage with large workspace, in *Proc. ASME Int. Design Engineering Technical Conf.*, Boston, MA, pp. 1–11.
[8] Wu, Y. and Zou, Q. (2009) Robust inversion-based 2-DOF control design for output tracking: Piezoelectric-actuator example. *IEEE Trans. Control Syst. Technol.*, **17** (5), 1069–1082.
[9] Xu, Q. (2015) Digital sliding mode prediction control of piezoelectric micro/nanopositioning system. *IEEE Trans. Contr. Syst. Technol.*, **23** (1), 297–304.

Part Two

Multi-Stroke Translational Micropositioning Systems

5

Dual-Stroke Uniaxial Flexure Stage

Abstract: This chapter presents the design and development of a flexure-based dual-stroke uniaxial micropositioning system. A coarse voice coil motor (VCM) and a fine piezoelectric stack actuator (PSA) are adopted to provide a long stroke and quick response, respectively. A decoupling design is carried out to minimize the interference behavior between the coarse and fine stages, by taking into account the actuation schemes and guiding mechanism implementations. Analytical model and finite-element analysis (FEA) results show that the system is capable of over 10 mm traveling while possessing a compact structure. A single-input/single-output (SISO) control scheme is realized on a prototype to demonstrate the performance of the dual-servo system (DSS). Experimental results confirm the superiority of the dual-servo stage over the stand-alone coarse stage and reveal the effectiveness of the presented decoupling design idea.

Keywords: Uniaxial micropositioning, Compliant mechanisms, Dual-stroke flexure mechanisms, Dual-servo system, Translational stages, Decoupling design, Finite-element analysis, Motion control.

5.1 Introduction

DSSs are highly desirable in micro-/nanomanipulation when high positioning accuracy, large motion range, and high servo bandwidth are required simultaneously. For instance, an atomic force microscope (AFM) usually has a scanning range less than 200 μm × 200 μm. In order to acquire the surface topology of a large specimen (e.g., over 10 mm × 10 mm), a nanopositioning stage with both a large workspace and a high bandwidth is required to fully cover the specimen surface and to quickly acquire the surface profile for areas of interest, as illustrated in Fig. 1.12. The objective of this chapter is to present the design methodology of a monolithic dual-servo flexure stage with a stroke greater than 10 mm along with a compact dimension for micropositioning applications.

To cater for the requirements of long stroke and high accuracy, a VCM and a PSA are employed for the coarse and fine drives of the DSS, respectively. Nevertheless, a DSS exhibits

Design and Implementation of Large-Range Compliant Micropositioning Systems, First Edition. Qingsong Xu.
© 2016 John Wiley & Sons Singapore Pte Ltd. Published 2016 by John Wiley & Sons Singapore Pte Ltd.

interference behavior (i.e., interaction between the coarse and fine actuators/stages). Based on the sensory information available for the output position, a DSS can be modeled as a dual-input/single-output (DISO) or a dual-input/dual-output (DIDO) system, which are special cases of a multiple-input/multiple-output (MIMO) system. Thus, it is natural to design a DSS controller by employing MIMO control techniques [1]. Although the MIMO control strategy is capable of handling the interference behavior explicitly, it usually produces a controller of very high order, which complicates its practical implementation in real-time control [2]. Therefore, some approaches have been proposed to simplify the DISO or DIDO plant by using SISO control schemes, such as a master–slave architecture [3], PQ method [4], or sensitivity decoupling approach [5].

Generally, the SISO techniques can be applied based on the assumption that the interference behavior in the DSS is neglected. However, such an assumption does not always hold in practice. For instance, it has been shown [6] that the open-loop system of the fine stage is unstable due to the severe interference behavior of the DSS. An impact force controller is employed [7] to control the fine stage so as to overcome the interference behavior. Yet, this control strategy requires excessive hardware because two fine actuators are used to realize the forward and backward actuation. Thus, it is desirable for designing a DSS properly by minimizing the interference behavior in order to apply the SISO control technique.

In the literature, some general guidelines have been presented for designing a DSS. For example, by analyzing the dynamics equation of the DSS, it has been suggested that the weight of the fine stage and the stiffness of the coarse stage should be minimized to reduce the interference behavior [8]. Also, a large stiffness of the fine stage has been reported [9, 10] to neglect the coupling effect in a DSS. Nevertheless, the available suggestions are inadequate to design a specific DSS with minimized interference. To this end, more comprehensive design guidelines are proposed in this chapter to design a DSS with minimal interference behavior.

Specifically, the design of the actuation scheme as well as the guiding mechanism of the fine actuator is proposed to minimize the interference effect. A uniaxial DSS with minimal interference is developed as an illustration. The presented design guidelines are verified by employing FEA simulations. Moreover, a SISO control scheme is realized in experimental studies to demonstrate the performance of the DSS without considering the interference behavior, which achieves a positioning resolution of 500 nm within the motion stroke over 10 mm along with a more rapid transient response than for the stand-alone coarse stage.

5.2 Mechanism Design and Analysis

In this section, some mechanism design considerations are proposed to minimize the interference behavior and to achieve a large stroke for the DSS.

5.2.1 Mechanism Design to Minimize Interference Behavior

The interference behavior of a DSS means the interaction between the coarse and fine actuators or the coarse and fine stages. To facilitate the control system design based on SISO control techniques, it is desirable to minimize the interference behavior.

5.2.1.1   Design of Actuation Scheme

According to the actuation schemes, DSSs can be classified into two categories in terms of dual-actuation and dual-stage types. The former [7, 11] implies that a single common stage is driven by a coarse and a fine actuator, and the two actuators are connected in series. The latter [12, 13] indicates that two stages are driven by a coarse and a fine actuator, respectively, and these two stages are then connected in series (e.g., in a stacked or nested manner).

For illustration, a dual-actuation micropositioning stage is depicted in Fig. 5.1(a). The single stage is driven by the coarse VCM and fine PSA, which are connected in series. A dual-stage micropositioning stage is shown in Fig. 5.1(b), where the PSA-driven fine stage is nested inside the VCM-driven coarse stage and the fine stage consists of two parallelogram flexures. The output displacement of the PSA is guided by a flexure mechanism, which is composed of eight right-circular hinges.

Concerning a dual-actuation DSS, the mass m_1 (including the moving coil of the VCM, one half of the PSA, and one half of the PSA guiding mechanism) is depicted in Fig. 5.1(a) and m_2 is the mass of the remaining part (including one half of the PSA, one half of the PSA guiding mechanism, and the coarse stage). For a dual-stage DSS, the mass m_2 (including one half of the PSA, one half of the PSA guiding mechanism, and the fine stage) is depicted in Fig. 5.1(b) and the mass of the remaining part (including one half of the PSA, one half of the PSA guiding mechanism, the coarse stage, and the moving coil of the VCM) is represented by m_1. It is observed that the conditions $m_1 < m_2$ and $m_2 < m_1$ generally hold for a dual-actuation and a dual-stage DSS, respectively.

To design a suitable actuation scheme, the dynamics model of a DSS is established by referring to a two-mass physical model as shown in Fig. 5.2. Parameters k_1 and k_2 represent the effective stiffnesses of the coarse and fine stages, respectively. In addition, c_1 and c_2 denote the damping coefficients of the coarse and fine stages, respectively. For the VCM, only the

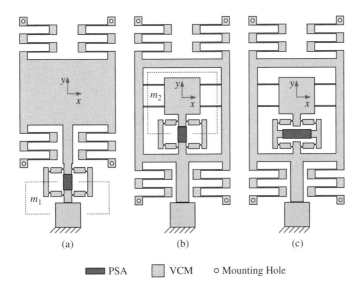

PSA   VCM   ○ Mounting Hole

Figure 5.1   Schematic diagrams of (a) dual-actuation and (b), (c) dual-stage micropositioning systems.

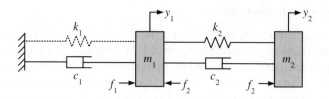

Figure 5.2 Mechanical model of a DSS.

damping coefficient c_1 is considered due to the nature of the back electromotive force [7]. By applying Newton's second law, the differential equations can be derived as

$$m_1\ddot{y}_1 + (c_1 + c_2)\dot{y}_1 - c_2\dot{y}_2 + k_2y_1 - k_2y_2 = f_1 - f_2 \tag{5.1}$$

$$m_2\ddot{y}_2 - c_2\dot{y}_1 + c_2\dot{y}_2 - k_2y_1 + k_2y_2 = f_2 \tag{5.2}$$

where y_1 and y_2 are the displacements of the coarse and fine stages, and f_1 and f_2 denote the actuation forces created by VCM and PSA actuators/stages, respectively.

By taking the Laplace transform of the dynamics equations, one can deduce that

$$Y_1(s) = q_{11}F_1(s) + q_{12}F_2(s) \tag{5.3}$$

$$Y_2(s) = q_{21}F_1(s) + q_{22}F_2(s) \tag{5.4}$$

where $Y_i(s)$ and $F_i(s)$ (for $i = 1$ and 2) represent the Laplace transforms of the displacement and force signals, respectively. Moreover, the four coefficients take on the forms

$$q_{11} = \frac{m_2s^2 + c_2s + k_2}{n_0} \tag{5.5}$$

$$q_{12} = \frac{-m_2s^2}{n_0} \tag{5.6}$$

$$q_{21} = \frac{c_2s + k_2}{n_0} \tag{5.7}$$

$$q_{22} = \frac{m_1s^2 + c_1s}{n_0} \tag{5.8}$$

where

$$n_0 = (m_1s^2 + c_1s + k_1)(m_2s^2 + c_2s) + m_2s^2(c_2s + k_2). \tag{5.9}$$

Generally, to minimize the interference behavior, the coefficients q_{12} and q_{21} should be minimized. In view of the numerators, it can be deduced that a smaller m_2 and c_2 and a smaller k_2 are desirable. For both actuation schemes of the DSS, the PSA has a smaller damping coefficient c_2 and a higher stiffness k_2. However, in the case of a dual-actuation DSS, the relationship $m_2 > m_1$ holds. That is, there always exists an interference effect. Therefore, a dual-stage scheme (i.e., $m_2 < m_1$) is designed for the DSS in order to minimize the interference behavior.

5.2.1.2   Guiding Mechanism Design of Fine Actuator

It is known that the PSA cannot bear large transverse loads due to the risk of damage. Thus, a suitable guiding mechanism is required for PSA actuation to isolate the transverse load. Figure 5.1(a) and (b) presents a motion guiding mechanism with an embedded PSA. The guiding mechanism is driven directly by the nested PSA, as shown in Fig. 5.3(a). Referring to the one-quarter mechanism as shown in Fig. 5.3(b), the input and output displacements are assumed to be Δx and Δy, respectively. Assume that the compliances of the right-circular hinges come mainly from the translational (Δl) and rotational ($\Delta \alpha$) motion of the hinges. In addition, let K_t and K_r denote the translational and rotational stiffness of one hinge, respectively. Then, based on the virtual work principle, it can be deduced that

$$\frac{f_{PSA}}{2}\Delta x - \frac{f_2}{2}\Delta y = 2K_r \Delta \alpha^2 \tag{5.10}$$

where f_{PSA} is the actuation force of the PSA and f_2 represents the external force applied at the output end of the amplifier. In view of $\Delta y = \Delta x$, it can be derived that

$$f_2 = f_{PSA} - \frac{4K_r \Delta \alpha^2}{\Delta x}. \tag{5.11}$$

It is observed from Eq. (5.3) that, in order to further reduce the interaction of a fine stage on the coarse stage for a dual-stage DSS, the force f_2 acting on the coarse stage should be alleviated. For such purpose, a bridge-type displacement amplifier is adopted as a guiding mechanism due to its compactness, as shown in Fig. 5.1(c). Close-up views of the amplifier and the right-circular hinge are shown in Fig. 5.3(c) and (d), respectively. The employed displacement amplifier acts as a displacement guiding and amplification device as well as

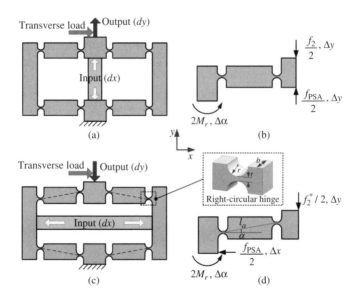

Figure 5.3   (a), (b) Parameters of the guiding mechanism; (c), (d) amplification principle and parameters of the displacement amplifier.

an interaction force reducer. Once driven by an input displacement, the device produces an amplified vertical output displacement along the backward direction.

Again, based on the virtual work principle, the following relation is deduced by referring to the one-quarter amplifier as shown in Fig. 5.3(d):

$$\frac{f_{PSA}}{2}\Delta x - \frac{f_2^*}{2}\Delta y = K_t\Delta l^2 + 2K_r\Delta\alpha^2. \tag{5.12}$$

In view of the amplification ratio ($A_a > 1$) of the displacement amplifier, it is derived that $\Delta y = A_a\Delta x$. Thus, Eq. (5.12) can be rewritten as

$$f_2^* = \frac{1}{A_a}f_{PSA} - \frac{2K_t\Delta l^2}{A_a\Delta x} - \frac{4K_r\Delta\alpha^2}{A_a\Delta x} \tag{5.13}$$

which indicates that the force exerted on the output end of the amplifier is reduced to about $\frac{1}{A_a}$ of the former result, as shown in Eq. (5.11). It follows that the interaction force between the fine and coarse stages has been reduced by about A_a times using the bridge-type displacement amplifier.

5.2.1.3   Amplification Ratio and Input Stiffness Calculation

Without considering the external force (f_2) exerted by the parallelogram fine stage, the amplification ratio and actuation stiffness of the amplifier can be calculated as follows [14]:

$$A_a = \frac{K_t l_a^2 \cos^3\alpha \, \sin\alpha}{2K_r + K_t l_a^2 \cos^2\alpha \, \sin^2\alpha} \tag{5.14}$$

$$K_{in} = \frac{2K_t K_r \cos^2\alpha}{2K_r + K_t l_a^2 \cos^2\alpha \, \sin^2\alpha} \tag{5.15}$$

where the amplifier parameters α and l_a are shown in Fig. 5.3(d). Concerning the stiffnesses K_r and K_t, the equations with the best accuracy as suggested in [15] are adopted in the current work.

Generally, the bridge-type displacement amplifier has a large transverse stiffness. In practice, the output end of the amplifier is connected to the output platform for the actuation. The high transverse stiffness of the amplifier is useful to tolerate the external load even if a large transverse load, as shown in Fig. 5.3(c), is exerted on the amplifier.

In addition, with reference to Fig. 5.4, the stiffness of the parallelogram fine stage can be derived as.

$$K_{fs} = \frac{4Ebh_2^3}{l_2^3} \tag{5.16}$$

where E is the Young's modulus of the material, h_2 and l_2 are the in-plane width and length of one leaf flexure, respectively.

To guarantee proper operation of the amplification device, the selected PSA should be powerful enough to drive the amplifier connected to the fine stage. Therefore, the stiffness of the PSA should satisfy the condition

$$K_{PSA} \geq K_{in} + K_{fs}. \tag{5.17}$$

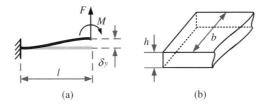

Figure 5.4 Deformation and parameters of a leaf-spring flexure.

5.2.2 Mechanism Design to Achieve Large Stroke

To achieve a large motion range, folded leaf springs composed of N flexures are used. In this research, $N = 4$ is adopted to illustrate the conceptual design, although more or fewer flexures may also be employed for the design.

5.2.2.1 Displacement and Stiffness Analysis

The dimensions of the leaf springs need to be well designed to ensure a large range of output motion without fatigue of the material. Referring to Fig. 5.1(b), each flexure in the four folded leaf springs suffers from identical deformation. The deformed shape of one flexure hinge is shown in Fig. 5.4. It is observed that each flexure bears a combined force F and moment M. In view of the boundary conditions in terms of rotational angle and translational displacement as shown in Eqs. (2.1) and (2.2), it can be deduced that

$$F = \frac{2M}{l_1}, \quad \delta y = \frac{Fl_1^3}{12EI} \tag{5.18}$$

where δy denotes the transverse displacement of one flexure, and $I = bh^3/12$ is the area moment of inertia of the cross-section with respect to the neutral axis.

Then, the stiffness of one leaf flexure observed at the output end can be derived as

$$K = \frac{F}{\delta y} = \frac{Ebh_1^3}{l_1^3}. \tag{5.19}$$

When the maximum moment M_{max} is exerted by the flexures, the maximum stress σ_{max} occurs at the outermost edge of the cross-section. The maximum stress can be calculated as

$$\sigma_{max} = \frac{M_{max}h_1}{2I} \tag{5.20}$$

which gives

$$M_{max} = \frac{\sigma_{max}bh_1^2}{6}. \tag{5.21}$$

Considering that $F_{max} = K\delta y_{max}$ and taking into account Eqs. (5.18), (5.20), and (5.21), the maximum translation for one leaf flexure is calculated as follows:

$$\delta y_{max} = \frac{F_{max}}{K} = \frac{2M_{max}}{Kl_1} = \frac{\sigma_{max}l_1^2}{3Eh_1} \tag{5.22}$$

which indicates that the maximum one-sided translation δy_{max} of the leaf flexure is governed by the length l and width h of the leaf flexures for a given material. Thus, to obtain a larger δy, longer and thinner flexures are desirable.

Considering that each folded leaf spring consists of N flexures connected in series, the maximum translation of the coarse stage can be derived as

$$\Delta y_{max} = N\delta y_{max} = \frac{N\sigma_{max} l_1^2}{3Eh_1}. \tag{5.23}$$

By comparing Eq. (5.23) with Eq. (5.22), the effect of the folded springs is clear. That is, they are used to increase the translation by N times compared with a single leaf flexure.

Assume that an external force F_y is applied on the stage's output platform, which induces a displacement Δy of the stage in the y-axis direction. Then, the stage stiffness observed at the output end can be derived as

$$K_{stage} = \frac{F_y}{\Delta y} = \frac{4F}{4\delta y} = \frac{F}{\delta y} = \frac{Ebh_1^3}{l_1^3}. \tag{5.24}$$

Because the PSA is nested inside the displacement amplifier and the stiffness of the amplifier seen at the output end is assumed to be infinite, the stiffness of the dual-actuation stage in the y-axis can be derived as

$$K_y = K_{stage} = \frac{Ebh_1^3}{l_1^3}. \tag{5.25}$$

5.2.2.2 Actuation Issue Consideration

As far as the coarse actuation is concerned, a VCM is selected to drive the positioning stage to generate a centimeter range of motion. However, the VCM exhibits a lower blocking force compared with the PSA. To facilitate VCM actuation, the stage should be designed with sufficiently low stiffness.

Given the required maximum one-sided displacement D_{max} of the stage in the output (y-axis) direction, the required maximum actuation force can be determined as

$$F_{actuation} = K_y D_{max}. \tag{5.26}$$

To select the motors, the VCM should be chosen such that the maximum actuation force F_{VCM} satisfies the condition

$$F_{VCM} \geq F_{actuation} = \frac{Ebh_1^3 D_{max}}{l_1^3}. \tag{5.27}$$

The above condition guarantees that the selected motor is powerful enough to overcome the elastic energy of the coarse positioning stage.

Table 5.1 Main parameters of a dual-stroke micropositioning stage

Parameter	Symbol	Value	Unit
Length of leaf flexure #1	l_1	22.0	mm
In-plane width of leaf flexure #1	h_1	0.7	mm
Length of leaf flexure #2	l_2	19.0	mm
In-plane width of leaf flexure #2	h_2	0.6	mm
Out-of-plane thickness of plate material	b	10.0	mm
Radius of right-circular hinge	r	2.0	mm
Minimum width of right-circular hinge	t	0.6	mm
Length of amplifier flexure	l_a	12.1	mm
Inclination angle of amplifier flexure	α	7.1	°

5.2.3 FEA Simulation and Design Improvement

In the simulation study, the alloy material Al 7075 is assigned for the stage. Without loss of generality, the main kinematic parameters of a dual-stage micropositioning stage are shown in Table 5.1, where subscripts 1 and 2 indicate the parameters for the coarse and fine stages, respectively. In addition, the material specifications are described in Table 3.2. Both statics and dynamics performances of the stage are verified by conducting FEA simulations.

Analytical model results reveal that the maximum allowable one-sided displacement of the stage is $\Delta y_{max} = 6.47$ mm. Thus, the motion range in the working axis is $\pm \Delta y_{max} = \pm 6.47$ mm. By assigning the motion range as ± 5 mm, the maximum force needed to drive the stage is $F_{max} = 115.48$ N. In addition, the amplification ratio and actuation stiffness of the displacement amplifier are calculated as $A_a = 7.65$ and $K_{in} = 8.00$ N/μm, respectively.

5.2.3.1 Statics Performance Test

In order to assess the statics performance of the stage, static structural FEA simulation is carried out by applying an input displacement at the input end. Specifically, using a 5-mm displacement input for the VCM and 14.5-μm displacement input for the PSA, the FEA results are illustrated in Fig. 5.5(a) and (b), respectively. Note that the coarse stage is left free when the fine stage is actuated, as shown in Fig. 5.5(b). Due to a nice property of the motion decoupling, the coarse stage remains almost stationary. Thus, no cross-coupled motion is observed for the coarse stage as shown in Fig. 5.5(b).

It is found that the required actuation force is 113.36 N for the VCM, and the maximum stress is achieved at 405.14 MPa. By adopting the yield strength as the maximum allowable stress of the material, the safety factor is calculated as 503 MPa/405.14 MPa = 1.24. In addition, PSA actuation with the maximum input displacement reveals a safety factor of 6.86. Considering the FEA result as the benchmark, it is observed that the analytical model overestimates the force requirement and maximum allowable displacement by 1.9% and 4.2%, respectively.

In addition, the FEA reveals that the amplification ratio of the displacement amplifier is 5.60, along with an input stiffness of 8.74 N/μm. Thus, in comparison with FEA results, the analytical model overestimates the amplification ratio by 36.6% and underestimates the input stiffness

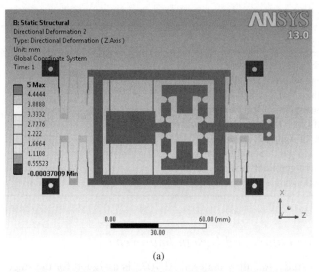

(a)

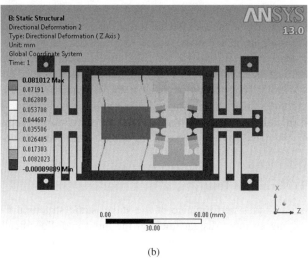

(b)

Figure 5.5   FEA results of the initial design: (a) VCM and (b) PSA driven, respectively. Only the four corner mounting holes are fixed, and the coarse stage is left free in (b).

by 8.5%. The relatively large discrepancy mainly comes from the assumptions employed in the analytical models, where only the bending compliances of flexure hinges are considered. The model accuracy can be improved by considering full compliance of the flexure mechanism.

5.2.3.2   Interference Testing Results

In order to examine the interference between the coarse and fine stages for the two dual-stage designs shown in Fig. 5.1(b) and (c), an identical actuation force is applied on the input end of the two fine stages. Note that only the four mounting holes are fixed, and the coarse stage is left free during the simulation. To characterize the interference behavior, the y-axis displacement of the input end of the coarse stage is defined as the interference motion caused by the fine stage.

By applying an input force of 10 N, interference motions of 0.5160 µm and 0.0648 µm are obtained for the two designs as shown in Fig. 5.1(b) and (c), respectively. Thus, the interference motion of the improved design shown in Fig. 5.1(c) has been reduced to 1/7.96 of the former design given in Fig. 5.1(b). Based on Hooke's law, it follows that the interference force has been reduced to 1/7.96 of the former design as well. In addition, the simulation result agrees well with the analytical result ($1/A_a = 1/7.65$), which is predicted by Eq. (5.14).

Moreover, FEA simulation shows that the interference motion of the coarse stage accounts for only 0.0648/6.3626 × 100% ≈ 1% of the motion of the fine stage, which demonstrates a negligible interference behavior in the improved design.

5.2.3.3 Modal Analysis and Structure Design Improvement

To evaluate the dynamics performance of the designed stage, modal analysis is conducted for the dual-stage design as depicted in Fig. 5.1(c). The first six resonant mode frequencies are tabulated in Table 5.2, and the first six resonant mode shapes are shown in Fig. 5.6. It is found that the first mode is associated with the major motion (i.e., the translation along the y-axis working direction) and the second one is the translation in the x-axis transverse direction. Moreover, the first two mode frequencies are very close to each other, which indicates that the stage is prone to transverse motion under external forces due to a low transverse stiffness. Because the transverse motion is passive and cannot be controlled by the actuators, it poses an obstacle to the control system design. Thus, in order to improve the robustness of the stage motion along the major motion direction, an improved structure design is desired to enhance the transverse stiffness.

An inspection of the deformations shown in Fig. 5.5(a) indicates that the two passive moving stages, which are located at each outer side of the stage, translate the identical magnitude of displacement. Thus, they can be connected together with connecting links without influencing the motion property of the stage. A schematic view and CAD model of the improved design are shown in Fig. 5.8.

Concerning the improved design, the FEA results of the first six resonant mode frequencies are also shown in Table 5.2, and the first six resonant mode shapes are depicted in

Table 5.2 The first six mode frequencies of the initial and improved designs of the dual stage

Mode no.	Initial design (Hz)	Improved design (Hz)	Improvement (%)
1	72.61	70.39	−3.0
2	83.70	172.37	105.9
3	112.48	216.74	92.7
4	123.45	225.15	82.4
5	195.02	270.99	39.0
6	284.38	301.18	5.9

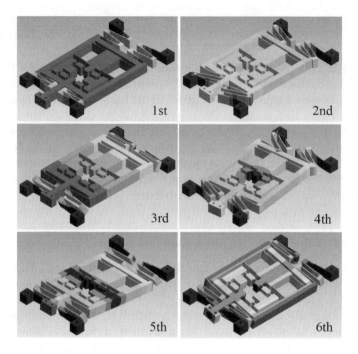

Figure 5.6   The first six resonant mode shapes of the initial dual-stage design.

Fig. 5.7. It is observed that the first and second modes are also associated with the y-axis major motion direction and x-axis transverse motion direction, respectively. Although the first mode frequency is slightly reduced by 3.0% due to an added mass of the connecting links, the second one is significantly increased by 105.9% compared with the original design. As a result, the second mode frequency is over twice as high as the first one, which reveals a robust motion along the major direction. Thus, from the control point of view, the improved design is preferable to the original design. It is notable that the amplification ratio and input stiffness of the fine stage are not changed, because only the design of the coarse stage is improved.

By inspecting the improved design as shown in Fig. 5.8, it is observed that the coarse stage is constructed by a mirror of two MCPFs with $N = 2$. It is notable that the secondary stages of the MCPFs can be linked together to generate a higher transverse stiffness in the x-axis direction.

5.3   Prototype Development and Open-Loop Testing

Using the architecture parameters as shown in Table 5.1, a stage prototype is fabricated from Al 7075 alloy by the wire-electrical discharge machining (EDM) process, which produces a compact dimension of 142 mm × 82 mm. To achieve a motion range of ±5 mm, FEA results suggest that the VCM actuation force and PSA stiffness should be chosen according to

$$F_{\text{VCM}} \geq 113.36 \text{ N}, \quad K_{\text{PSA}} \geq 8.74 \text{ N/}\mu\text{m}. \tag{5.28}$$

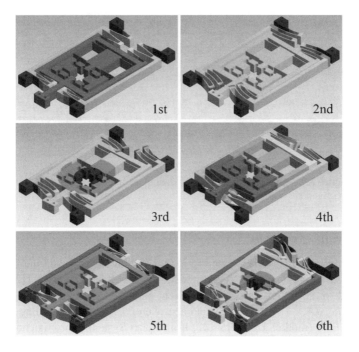

Figure 5.7 The first six resonant mode shapes of the improved dual-stage design.

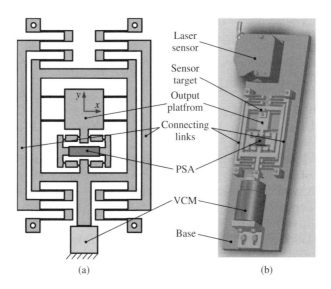

Figure 5.8 (a) Schematic diagram and (b) CAD model of the improved design of dual-stage micropositioning system.

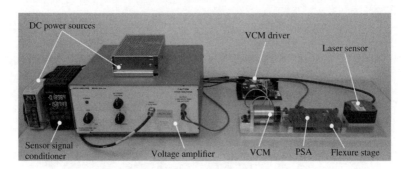

DC power sources

VCM driver

Laser sensor

Sensor signal
conditioner

Voltage amplifier   VCM   PSA   Flexure stage

Figure 5.9   Experimental setup of a dual-stage micropositioning system.

5.3.1   Experimental Setup

The fabricated prototype stage is shown in Fig. 5.9. Taking into account the force capability and motion range requirements, the VCM (model: NCC05-18-060-2X, from H2W Techniques, Inc.) is selected for actuation. Driven by a current amplifier, it provides a large enough output force of 27.8 N/A × 7 A = 194.6 N and a stroke of 12.7 mm. According to the stiffness requirement, a PSA (model: TS18-H5-202, from Piezo Systems, Inc.) is chosen. It offers a large stiffness of 230 N/μm and a stroke of 14.5 μm. Additionally, a high-voltage amplifier (model: EPA-104 from Piezo Systems, Inc.) is used to amplify the input voltage of ±10 V to ±200 V to drive the PSA. The stage output motion in the working axis is detected by a non-contact laser displacement sensor (model: LK-H055, from Keyence Corp.) with a measuring range of 20 mm. The controller is implemented with an NI USB-6259 board (from National Instruments Corp.) equipped with 16-bit D/A and A/D channels. The control algorithms are developed with LabVIEW software.

5.3.2   Statics Performance Testing

The open-loop statics performance of the micropositioning stage is tested to examine the stroke of the system. With 1-Hz sinusoidal voltage signals applied to VCM and PSA, the sensor readings are recorded as shown in Fig. 5.10(a) and (b), respectively. It is observed from Fig. 5.10 that the VCM coarse stage produces a motion stroke of 10.25 mm and the PSA fine stage delivers an output of 94.92 μm. Thus, the dual stage has an overall stroke of around 10.3 mm. The hysteresis effects mainly come from the adopted actuators, which can be suppressed by employing suitable control strategies.

The aforementioned motion range of the fine stage is obtained by leaving the VCM moving during the actuation of the PSA. In contrast, when the moving coil of the VCM is fixed at the base, the motion range of the PSA-driven fine stage is generated as 95.56 μm. This is slightly larger than the aforementioned value of 94.92 μm, which is obtained under the interaction of coarse and fine stages. Thus, the interference motion of the coarse stage induced by the fine stage can be calculated as 0.64 μm, which is less than 0.7% of the fine stage's motion. It is observed that the experimental result is consistent with the FEA simulation result of 1%.

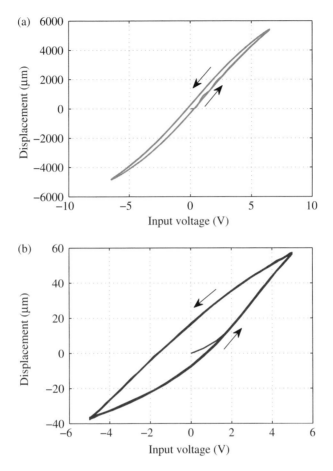

Figure 5.10 Open-loop output–input relations of the dual-stage micropositioning system: (a) VCM and (b) PSA driven, respectively.

In addition, the experimental result indicates an amplification ratio of 6.59 for the displacement amplifier, which is 17.7% higher than the simulation result. The discrepancy is mainly caused by the preloading effect when mounting the PSA.

5.3.3 Dynamics Performance Testing

The dynamics performance of the dual-stage system is tested by the frequency response method. The frequency responses of the stage are shown in Fig. 5.11(a) and (b), which are obtained with stand-alone VCM and PSA drive, respectively. With the two types of actuation, the resonant frequencies of 35 Hz and 495 Hz are identified. In comparison with the FEA simulation result of 70 Hz, the lower resonant frequency of the VCM stage is mainly induced by the added mass of the moving coil of the motor, which is not considered in FEA simulation. Additionally, the PSA actuation produces a much higher resonant frequency, which enables a quicker response than with VCM actuation.

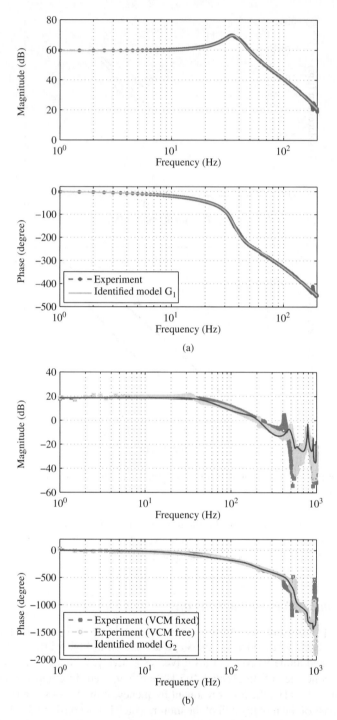

Figure 5.11   Open-loop frequency responses of the dual-stage micropositioning system: (a) VCM and (b) PSA driven, respectively.

The aforementioned frequency response of the fine stage is obtained by leaving the VCM free during the actuation of the PSA. When the moving coil of the VCM is fixed, the generated frequency response is also shown in Fig. 5.11(b), which indicates the first resonant frequency at 417 Hz. It is observed that the interference behavior increases the first resonant frequency by 18.7% compared with the result with the VCM fixed. In the frequency range of 0 to 300 Hz, the deviations of the magnitude and phase of the two frequency responses are within 2.2 dB and 19.3°, respectively. Thus, the interference behavior between the fine and coarse stages is negligible up to 300 Hz.

Due to the lack of accurate physical parameters of the actuators, the frequency responses are used to estimate the plant models of the system. As shown in Fig. 5.11, a 4th-order model G_1 and a 14th-order model G_2 are identified for the coarse and fine stages, respectively. The estimated linear models are employed for the control design in the following section.

5.4   Controller Design and Experimental Studies

The dual-servo control has been investigated extensively. In order to verify the presented design ideas of the dual-stage micropositioning system and to achieve a precise positioning, a global control scheme is realized and examined by conducting a series of experimental studies.

5.4.1   Controller Design

As shown in Fig. 5.12, a control scheme is implemented on the micropositioning system by taking the error of the coarse stage as the reference input to the fine stage. Because only one absolute position sensor of the fine stage is available, the relative position between the two stages is generated by employing an estimation technique. Different from the existing control scheme where an extra relative displacement sensor is used [13], the position output (y_k^1) of the coarse stage is offered by an observer as shown in Fig. 5.12. The role of the observer is to provide position information on the coarse stage by passing the input voltage (u_k^1) through the identified model G_1 of the coarse stage. After, the relative displacement (y_k^2) of the fine stage is generated by subtracting the coarse stage output from the sensor readings, i.e., $y_k^2 = y_k - y_k^1$.

In this work, the popular proportional-integral-derivative (PID) control is employed for both the coarse and fine stages. As is known, PID is a model-free controller which solves the control command by making use of the control error only. A digital PID control scheme can be realized

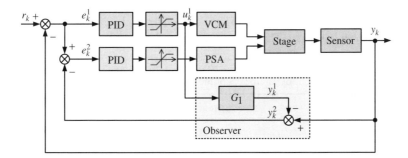

Figure 5.12   The control scheme for the dual-stage micropositioning system.

as follows:

$$u_k = K_p e_k + K_i \sum_{i=0}^{k} e_k + K_d(e_k - e_{k-1}) \tag{5.29}$$

with the positioning error $e_k = r_k - y_k$, where r_k and y_k represent the desired and actual system outputs at the k th time step, and K_p, K_i, and K_d denote the proportional, integral, and derivative gains, respectively. The key issue in PID design lies in the controller parameter tuning. Here, the Ziegler–Nichols (Z–N) method is adopted owing to its popularity.

It is notable that the position error of the coarse stage is used as the reference input to the fine stage. Initially, this error may be very large relative to the motion range of the PSA stage. Moreover, the PSA responds very quickly compared with the VCM. Consequently, there may be an over-demand for the PSA during a long duration until the position error of the coarse stage is kept within the motion range (i.e., 94 μm) of the PSA stage. To avoid possible destruction of the actuators, the saturation functions are used in both VCM and PSA control loops to limit the control action within ±6.5 V and ±5 V, respectively.

5.4.2 Experimental Studies

With the designed global control scheme, experimental studies are conducted to verify the performance of the micropositioning system. A sampling time of 5 ms is employed to implement the real-time controller. By experiment, the PID control parameters are tuned with the Z–N method to eliminate the overshoot. The PID control gains of the coarse and fine stages are tuned as shown in Table 5.3.

For illustration, a 200-μm set-point positioning is carried out. The experimental results of the coarse actuation and dual actuation are shown in Fig. 5.13(a), and the positioning errors are shown in Fig. 5.13(b). The histograms of the steady-state errors are compared in Fig. 5.14. With VCM actuation alone, a 5% settling time of 0.495 s and a steady-state root-mean-square error (RMSE) of 0.306 μm is produced. By using the dual actuation of VCM and PSA, the settling time is reduced to 0.357 s and the RMSE is suppressed to 0.257 μm. Compared with the stand-alone coarse stage, the dual-stage scheme has substantially improved the transient response speed by 28% and reduced the steady-state error by 16%, which results in a significant increase of the bandwidth along with a positioning resolution of 500 nm. It is notable that the Z–N method does not produce optimal parameters for the PID controller. The steady-state error can be further reduced by optimally tuning the PID controllers or employing displacement sensors with finer resolution.

In addition, for the dual-stage micropositioning system, the components of the coarse and fine control actions are shown in Fig. 5.15. It is observed that the PSA-driven fine stage takes effect mainly at the initial states. Then, its control action decays gradually to a constant value as

Table 5.3 PID control parameters of the coarse and fine stages

Control parameter	Coarse stage	Fine stage
K_p	1.28×10^{-4}	0.04
K_i	0.0085	3.36
K_d	4.80×10^{-7}	1.31×10^{-4}

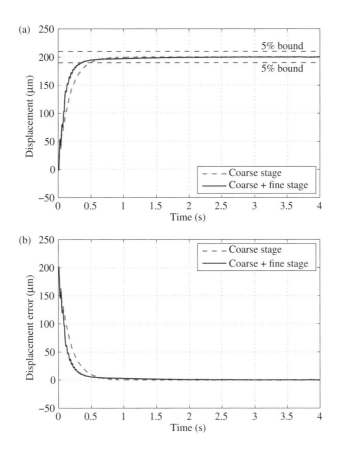

Figure 5.13 The 200-µm set-point positioning results of the coarse stage and dual-stage systems: (a) positioning results; (b) positioning errors.

time elapses. As a result of the fine actuation, the transient response speed is improved and the positioning error is gradually suppressed towards zero. The experimental results demonstrate the effectiveness of the control scheme for the dual-stage servo system.

Because the control scheme has been implemented without considering the interaction behavior between the coarse and fine stages, the experimental results reveal the effectiveness of the proposed ideas for the decoupling design of the dual-stage system. Considering that the plant model G_1 of the coarse stage is used as an observer to obtain the coarse stage output in the control scheme, the results also indicate the accuracy of the identified plant model. It is notable that the hysteretic nonlinearity of the PSA is not treated explicitly, in that the hysteresis effect over a small portion of stroke is not significant. Hence, it is adequately suppressed by the feedback control. In the future, a more effective control scheme will be employed to further improve the performance of the dual-stage micropositioning system.

5.5 Conclusion

A dual-stage micropositioning stage driven by a VCM and a PSA is presented in this chapter. New ideas on structure design are proposed to minimize the interaction between the two stages.

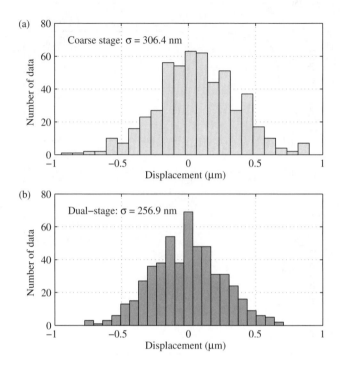

Figure 5.14   Histograms of steady-state errors of the coarse stage and dual-stage systems.

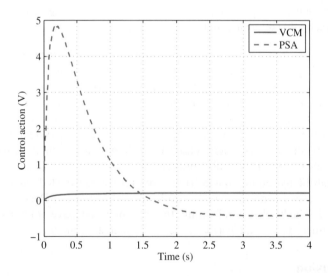

Figure 5.15   Control action components of the dual-stage system.

The decoupling design is verified by both FEM simulations and experimental studies. A global control scheme is employed to realize precision positioning. The proposed decoupling design allows the employment of a simple control scheme, since SISO control is applicable to both coarse and fine stages. Experimental results reveal that the dual-actuation scheme is superior to the stand-alone coarse-actuation one in terms of transient response and steady-state response. A quick positioning with a 500-nm resolution and over 10-mm stroke has been achieved by the coarse–fine cooperative system, which demonstrates the potential of the reported dual-stage system in the field of micro-/nanopositioning. In the future, more sophisticated control strategies will be exploited to further improve the positioning performance of the system.

References

[1] Suh, S.M., Chung, C.C., and Lee, S.H. (2002) Design and analysis of dual-stage servo system for high track density HDDs. *Microsyst. Technol.*, **8** (2&3), 161–168.

[2] Al Mamun, A., Mareels, I., Lee, T.H., and Tay, A. (2003) Dual stage actuator control in hard disk drive – a review, in *Proc. 29th Annual IEEE Industrial Electronics Society Conf.*, pp. 2132–2137.

[3] Chen, X., Zhang, S., Bao, X., and Zhao, H. (2008) Master and slave control of a dual-stage for precision positioning, in *Proc. 3rd IEEE Int. Conf. on Nano/Micro Engineered and Molecular Systems*, pp. 583–587.

[4] Rapley, H.R. and Messner, W.C. (2001) Designing controllers for two stage disk drive actuator systems using the PQ method and the sbode plot. *IEEE Trans. Magnetics*, **37** (2), 944–948.

[5] Horowitz, R., Li, Y., Oldham, K., Kon, S., and Huang, X. (2007) Dual-stage servo systems and vibration compensation in computer hard disk drives. *Control Eng. Practice*, **15** (3), 291–305.

[6] Song, Y., Wang, J., Yang, K., Yin, W., and Zhu, Y. (2010) A dual-stage control system for high-speed, ultra-precise linear motion. *Int. J. Adv. Manuf. Technol.*, **48**, 633–643.

[7] Liu, Y.T., Fung, R.F., and Wang, C.C. (2005) Precision position control using combined piezo-VCM actuators. *Precis. Eng.*, **29**, 411–422.

[8] Choi, H.S., Song, C.W., Han, C.S., Choi, T.H., Lee, N.K., Lee, H.W., *et al.* (2003) Designing compensator of dual servo system for high precision position control, in *Proc. SICE Annual Conf.*, pp. 1650–1655.

[9] Liu, Y., Li, T., and Sun, L. (2009) Design of a control system for a macro–micro dual-drive high acceleration high precision positioning stage for IC packaging. *Sci. China Ser. E-Tech. Sci.*, **52** (7), 1858–1865.

[10] Zheng, J., Su, W., and Fu, M. (2010) Dual-stage actuator control design using a doubly coprime factorization approach. *IEEE/ASME Trans. Mechatronics*, **15** (3), 339–348.

[11] Elfizy, A.T., Bone, G.M., and Elbestawi, M.A. (2005) Design and control of a dual-stage feed drive. *Int. J. Mach. Tools Manuf.*, **45**, 153–165.

[12] Michellod, Y., Mullhaupt, P., and Gillet, D. (2006) Strategy for the control of a dual-stage nano-positioning system with a single metrology, in *Proc. IEEE Conf. on Robotics, Automation and Mechatronics*, pp. 1–8.

[13] Dong, W., Tang, J., and ElDeeb, Y. (2009) Design of a linear-motion dual-stage actuation system for precision control. *Smart Mater. Struc.*, **18**, 095 035–1–095 035–11.

[14] Xu, Q. and Li, Y. (2011) Analytical modeling, optimization and testing of a compound bridge-type compliant displacement amplifier. *Mech. Mach. Theory*, **46** (2), 183–200.

[15] Yong, Y.K., Lu, T.F., and Handley, D.C. (2008) Review of circular flexure hinge design equations and derivation of empirical formulations. *Precis. Eng.*, **32** (2), 63–70.

6

Dual-Stroke, Dual-Resolution Uniaxial Flexure Stage

Abstract: This chapter presents the design and testing of a novel dual-stroke, dual-resolution precision positioning stage driven by a single linear actuator. The variable-stiffness structure is devised with leaf flexures to achieve a large stroke. Strain-gauge sensors are employed to provide different resolutions in the two motion ranges. To quantify the design of the motion ranges and the coarse/fine resolution ratio, analytical models are established and verified through finite-element analysis (FEA) simulations. A proof-of-concept prototype is fabricated for experimental studies and the results validate the effectiveness of the presented design.

Keywords: Uniaxial micropositioning, Compliant mechanisms, Flexure mechanisms, Translational stages, Dual-stroke stages, Dual-range design, Dual-resolution design, Finite-element analysis, Strain-gauge sensors, Variable-stiffness mechanism.

6.1 Introduction

Dual-range stages are demanded in precision positioning applications that call for fine resolution in a smaller motion range and coarse resolution in a larger range. The major issue of a dual-servo stage arises from the interference caused by the interaction between the coarse and fine stages. It has been shown that the interference behavior of a dual-servo stage can lead to an unstable open-loop control system [1]. To mitigate this adverse interference effect, control and mechanical design approaches have been developed. For instance, the interference behavior can be reduced by designing a multiple-input/multiple-output (MIMO) control system [2, 3]. The interaction effect can also be alleviated by resorting to a decoupling mechanical design as shown in Chapter 5. Even so, the employment of two types of actuators complicates the control and mechanism design processes [4].

Generally, one actuator can only deliver a single stroke along with a specific resolution. It is challenging to devise a single-drive stage with multiple strokes as well as multiple resolutions. In this work, the concept of a variable-stiffness mechanism is employed to accomplish the objective. A single-drive micropositioning stage with dual ranges and dual resolutions is

Design and Implementation of Large-Range Compliant Micropositioning Systems, First Edition. Qingsong Xu.
© 2016 John Wiley & Sons Singapore Pte Ltd. Published 2016 by John Wiley & Sons Singapore Pte Ltd.

devised for illustration. In particular, a conceptual design for a dual-range, dual-resolution compliant stage is proposed based on an unequal-stiffness compliant mechanism. In the smaller and larger ranges, the mechanism is designed to exhibit different stiffnesses, which are contributed by the leaf flexures experiencing different bending deformations. The larger and smaller deformations are monitored using two strain-type sensors to offer fine and coarse resolutions for the two ranges, respectively.

Rather than using dual-servo stages, the presented technique enables the achievement of a dual-range motion by adopting a single actuator. This reduces the hardware cost and control design effort. Moreover, the single-drive design eliminates the conventional interference effect. Both simulation and experimental investigations are conducted to verify the proof-of-concept design.

6.2 Conceptual Design

The conceptual design of a micropositioning stage with dual ranges and dual resolutions is presented in this section.

6.2.1 Design of a Compliant Stage with Dual Ranges

The schematic presentation of a uniaxial, dual-range micropositioning stage is depicted in Fig. 6.1. The output platform M is actuated by a linear actuator through a compliant guiding bearing #1, which has an equivalent stiffness K_1. In addition, the platform M is connected to a fixed base via a compliant guiding bearing #2, which exhibits a stiffness K_2. To yield two motion ranges for M, a mechanical mover is linked to the driving end of the actuator. The bidirectional translation of the mover is constrained by two mechanical stoppers, which are fixed on the output platform M. It is notable that the two stoppers and the output platform M move together. Without loss of generality, it is assumed that the initial clearances (δ) between the mover and the two stoppers are identical.

Referring to Fig. 6.1, once the actuator drives the output platform to move forward, both bearings #1 and #2 are compressed. The overall equivalent stiffness can be expressed as

$$K_{\text{range1}} = \frac{1}{\dfrac{1}{K_1} + \dfrac{1}{K_2}}. \tag{6.1}$$

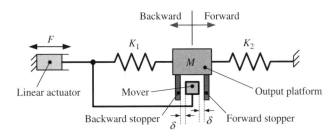

Figure 6.1 Schematic of a uniaxial micropositioning stage with dual ranges.

Assume that the relationship $K_1 < K_2$ holds. After a certain driving displacement D_1, the mover translates over a distance δ relative to M (i.e., the clearance between the mover and forward stopper). Then, it contacts the stopper. The corresponding displacement R_1 of the output platform can be calculated from the relationship

$$K_1(D_1 - R_1) = K_2 R_1 \tag{6.2}$$

which describes the driving force of the actuator. It allows the following generation:

$$R_1 = \frac{K_1}{K_1 + K_2} D_1. \tag{6.3}$$

Afterwards, if driving continues in the forward direction, only the bearing K_2 will be deformed because the deformation of the bearing K_1 is stopped by the forward stopper. Under this situation, the overall stiffness of the mechanism becomes

$$K_{\text{range2}} = K_2. \tag{6.4}$$

After the moment when the mover contacts the forward stopper, if a maximal driving displacement D_2 is produced by the actuator, then D_2 will be transmitted as the displacement of the output platform M. Hence, the overall output displacement of M can be derived as follows:

$$R_{\text{all}} = R_1 + R_2 = \frac{K_1}{K_1 + K_2} D_1 + D_2. \tag{6.5}$$

Therefore, the forward motion range of M is divided into two intervals $[0, R_1]$ and $[R_1, R_1 + R_2]$, which are assumed to be the smaller and larger ranges, respectively.

Similarly, the output platform can also be driven to move in the backward direction. Its backward motion range is divided into two intervals $[-R_1, 0]$ and $[-R_1 - R_2, -R_1]$ by the backward stopper.

Unlike the conventional variable-stiffness mechanism, which usually exhibits a specified stiffness profile [5, 6], the proposed design possesses two discrete stiffness values in the overall motion range. In the smaller and larger ranges, the equivalent stiffness of the system is unequal, although the stiffness remains constant in each range. Based on the foregoing dual-range design, a dual-resolution stage is devised in the next section.

6.2.2 Design of a Compliant Stage with Dual Resolutions

From the foregoing analysis, it is observed that in the smaller motion range $[-R_1, R_1]$, the deformation is experienced by both bearings, whereas in the larger motion ranges $[-R_1 - R_2, -R_1]$ and $[R_1, R_1 + R_2]$, the deformation of the mechanism is attributed to bearing #2 alone. In the smaller range, the deformations Δ_1 and Δ_2 of the bearings #1 and #2, respectively, are related by

$$K_1 \Delta_1 = K_2 \Delta_2 \tag{6.6}$$

which describes the driving force of the actuator.

Assume that $K_1 < K_2$, then it can be deduced from Eq. (6.6) that $\Delta_1 > \Delta_2$. That is, the deformation of bearing #1 is greater than that of bearing #2. It is known that strain-type sensors can be employed to measure the displacement of compliant mechanisms indirectly by detecting the varying strain of the deformed material [7]. If the same kind of strain sensor is adopted to measure the two different deformations, the larger the deformation is, the larger the output signal amplitude will be. That is, a larger deformation results in a higher signal-to-noise ratio (SNR) (i.e., higher measurement resolution).

Therefore, the deformation of bearing #1 can be monitored using a strain sensor to obtain a higher position resolution in the smaller range $[-R_1, R_1]$. By using the same type of sensor, the deformation of bearing #2 can be measured with a lower resolution in the larger ranges $[-R_1 - R_2, -R_1]$ and $[R_1, R_1 + R_2]$. In this way, a micropositioning stage with dual ranges and dual resolutions is devised. Specifically, the higher and lower resolutions are generated in the smaller and larger motion ranges, respectively.

6.3 Mechanism Design

An embodiment of the proposed compliant stage is designed as shown in Fig. 6.2. The leaf flexures are adopted to enable a large motion range of the compliant stage. Note that the mechanical stopper is mounted on the rear side of the planar mechanism. Several variations of the micropositioning stage with different fixing schemes are shown in Fig. 6.3.

Referring to Fig. 6.2, bearing #1 is denoted by its equivalent stiffness K_1, and the remaining flexures belong to bearing #2, which possesses an equivalent stiffness K_2. Bearing #2 is designed using MCPFs with $N = 2$ to generate a large motion range while maintaining the mechanism's compact size. A relatively large in-plane transverse stiffness is guaranteed by bearing #2. To produce a stage with the desired performance in terms of motion ranges and coarse/fine resolution ratio, parametric design of the flexure mechanism is conducted below.

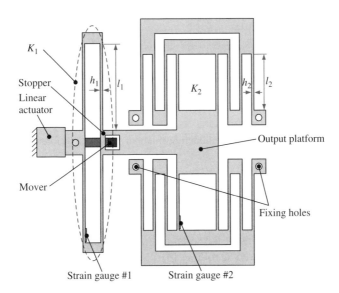

Figure 6.2   Schematic of a compliant micropositioning stage with dual ranges and dual resolutions.

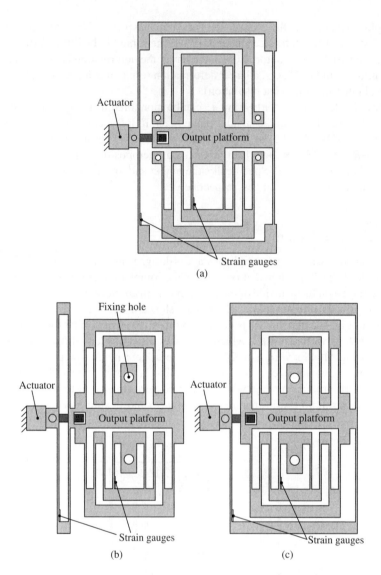

Figure 6.3   Variations of the uniaxial micropositioning stage with different fixing schemes.

6.3.1   Stiffness Calculation

The equivalent stiffnesses of the two guiding bearings are calculated analytically. Bearing #1 consists of four fixed–guided flexures, which experience identical deformation due to having the same dimensions. In view of the serial and parallel connections of these flexures, the equivalent stiffness can be derived as follows [8]:

$$K_1 = \frac{Ebh_1^3}{l_1^3} \tag{6.7}$$

where E is the Young's modulus of the material, and b, h_1, and l_1 represent the out-of-plane thickness, in-plane width, and length of the leaf flexures, respectively, as shown in Fig. 6.2.

In addition, considering that bearing #2 is composed of two MCPFs with $N = 2$, which are connected in parallel, its stiffness can be derived below in view of Eq. (2.11):

$$K_2 = \frac{Ebh_2^3}{l_2^3} \tag{6.8}$$

where h_2 and l_2 describe the in-plane width and length of the associated leaf flexures, respectively, as shown in Fig. 6.2.

6.3.2 Motion Range Design

Assume that the smaller and larger motion ranges are $[0, |R_1|]$ and $[|R_1|, |R_1 + R_2|]$, respectively. The following design shows how to generate these ranges.

6.3.2.1 Smaller Range Design

To produce a smaller motion range R_1, the absolute deformation of bearing #1 can be derived as follows:

$$\delta = D_1 - R_1 \tag{6.9}$$

where D_1 is the driving displacement of the actuator.

Then, substituting Eq. (6.3) into Eq. (6.9), a fundamental algebra operation gives

$$R_1 = \frac{K_1}{K_2}\delta \tag{6.10}$$

which indicates that the magnitude of the smaller motion range R_1 is governed by the clearance δ and the effective stiffnesses K_1 and K_2 of the two bearings.

Therefore, to obtain the smaller range R_1, the clearance between the mechanical mover and each stopper should be designed as

$$\delta = \frac{K_2}{K_1}R_1 \tag{6.11}$$

which is calculated from Eq. (6.10).

The required input displacement from the actuator can be obtained from Eq. (6.3):

$$D_1 = \left(1 + \frac{K_2}{K_1}\right)R_1. \tag{6.12}$$

Meanwhile, to avoid plastic deformation of the flexures, the stress induced by the deflection δ of bearing #1 should stay within the yield strength of the material. The allowable maximum motion range of bearing #1 (i.e., the allowable maximum clearance) can be computed as follows:

$$\delta^{\max} = \frac{2\sigma_y l_1^2}{3Eh_1} \tag{6.13}$$

where σ_y denotes the yield strength of the material.

To ensure the safety of the material, the magnitude of the smaller motion range should be designed as follows:

$$R_1 \le R_1^{\max} = \frac{K_1}{K_2}\delta^{\max} = \frac{2\sigma_y l_1^2 K_1}{3Eh_1 K_2}. \tag{6.14}$$

6.3.2.2   Larger Range Design

Taking into account that only bearing #2 is deformed in the larger motion interval R_2, the required driving displacement from the actuator is derived as

$$D_2 = R_2. \tag{6.15}$$

Similarly, to guarantee the safety of the material, the allowable maximum motion range of bearing #2 is calculated as follows:

$$\Delta_2^{\max} = \frac{4\sigma_y l_2^2}{3Eh_2}. \tag{6.16}$$

Therefore, the span of the larger motion range should be designed as follows:

$$R_2 \le R_2^{\max} = \Delta_2^{\max} - R_1 = \frac{4\sigma_y l_2^2}{3Eh_2} - R_1 \tag{6.17}$$

so that the safety of the material is guaranteed.

6.3.3   Motor Stroke and Driving Force Requirement

In this work, a VCM is employed to generate a relatively large motion range. To produce a one-sided entire motion range of $R_1 + R_2$, a driving displacement is needed:

$$D_1 + D_2 = \left(1 + \frac{K_2}{K_1}\right)R_1 + R_2. \tag{6.18}$$

The above displacement should not exceed the one-sided stroke D_{stroke} of the selected actuator, i.e.,

$$D_{\text{total}} = \left(1 + \frac{K_2}{K_1}\right)R_1 + R_2 \le D_{\text{stroke}}. \tag{6.19}$$

In addition, considering that the VCM typically delivers a not-large blocking force, the stage should be compliant enough so that the elastic energy can be overcome by the VCM. Assume that $K_1 < K_2$. Then it can be deduced from Eqs. (6.1) and (6.4) that $K_{\text{range1}} < K_{\text{range2}}$; that is, the stiffness in the smaller motion range is lower than that in the larger range. Thus, it can be deduced that the maximum force is needed to produce the extremum of the larger motion range. The required maximum driving force is calculated as follows:

$$F_{max} = K_{range2}(R_1 + R_2) \leq F_{actuator} \qquad (6.20)$$

where $F_{actuator}$ denotes the maximum driving force of the VCM actuator.

Substituting Eqs. (6.4) and (6.8) into Eq. (6.20) yields

$$F_{max} = \frac{Ebh_2^3}{l_2^3}(R_1 + R_2) \leq F_{actuator} \qquad (6.21)$$

which provides a guideline for the stage's parameter design.

6.3.4   Sensor Deployment

To measure the displacement of the compliant stage in the smaller and larger motion ranges, two sets of strain gauges are employed. The position of the maximum stress can be determined by conducting an FEA simulation. To enhance the SNR, the strain gauges are attached around the maximum-stress positions of the leaf flexures related to bearings #1 and #2, respectively, as depicted in Fig. 6.2.

The relationship described by Eq. (6.6) indicates that

$$\frac{\Delta_1}{\Delta_2} = \frac{K_2}{K_1}. \qquad (6.22)$$

By selecting $K_1 < K_2$, the relation $\Delta_1 > \Delta_2$ can be determined. It follows that in the smaller motion range, the length change value of strain gauge #1 is larger than that of strain gauge #2. Therefore, a higher SNR is expected for gauge #1. Hence, gauge #1 achieves a better displacement resolution than gauge #2.

Without loss of generality, assume that the output displacement of the stage is measured by the strain-gauge sensors through quarter-bridge circuits of signal conditioning, as shown in Fig. 6.4. The output voltage of the bridge circuit can be approximated by [9]

$$V_o = \frac{V_s}{4R} \times dR \qquad (6.23)$$

where dR and R represent the change value and nominal value of the gauge resistance, respectively.

The gauge factor can be expressed as

$$S = \frac{dR/R}{\varepsilon} \qquad (6.24)$$

where ε is the corresponding strain induced by the deformation of the flexure bearings. The strain ε is related to the experienced stress σ of the flexure by

$$\sigma = E\,\varepsilon \qquad (6.25)$$

where E is the Young's modulus of the material.

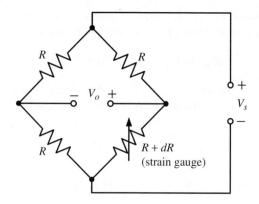

Figure 6.4   Wheatstone bridge circuit with a single strain gauge and three fixed resistors.

Given the mechanics of the material, the relationship between the stress σ and the guided deflection μ of a leaf flexure can be derived as

$$\sigma = \frac{Khl\mu}{4I} \tag{6.26}$$

where $K = Ebh^3/l^3$ and $I = bh^3/12$ are the linear stiffness and moment of inertia for a leaf flexure, respectively.

In view of the above equations, a relationship between the circuit output voltage V_o and the flexure deflection μ can be derived as follows:

$$\mu = \frac{4l^2 V_o}{3hSV_s}. \tag{6.27}$$

It is notable that the deflections of the flexures (μ_1 and μ_2) associated with bearings #1 and #2 are equal to a half and a quarter of the entire translations (Δ_1 and Δ_2) of the bearings, respectively. That is,

$$\mu_1 = \frac{\Delta_1}{2}, \quad \mu_2 = \frac{\Delta_2}{4}. \tag{6.28}$$

Hence, the ratio of output voltages of the two strain gauges is obtained as

$$\frac{V_{o1}}{V_{o2}} = \frac{2h_1 l_2^2 \Delta_1}{h_2 l_1^2 \Delta_2}. \tag{6.29}$$

Given Eq. (6.22), the above relation (6.29) becomes

$$\frac{V_{o1}}{V_{o2}} = \frac{2h_1 l_2^2 K_2}{h_2 l_1^2 K_1}. \tag{6.30}$$

In the following section, a dual-range, dual-resolution compliant stage is devised to illustrate the foregoing design procedures.

6.4 Performance Evaluation

Once the desired motion ranges R_1 and R_2 of the compliant stage are specified, the relationships involving Eqs. (6.13), (6.14), (6.17), (6.19), and (6.21) provide guidelines for the parameter design.

As a case study, a micropositioning stage is designed to produce the smaller and larger motion ranges of $R_1 = 0.2$ mm and $R_2 = 3.0$ mm, respectively. The adopted VCM provides a stroke of ± 5 mm and a maximum driving force of 29.2 N. The stage material is chosen as Al 7075 and the stage parameters are designed as shown in Table 6.1.

6.4.1 Analytical Model Results

The stiffnesses K_1 and K_2 of the two bearings are calculated using Eqs. (6.7) and (6.8), respectively. The results are summarized in Table 6.2. The analytical models predict that $R_1^{max} = 3.05$ mm, $R_2^{max} = 10.48$ mm, $D_{total} = 4.28$ mm, and $F_{max} = 12.30$ N, which all satisfy the aforementioned design criteria. In addition, the calculated total motion range of the stage is ± 3.2 mm.

The parametric design leads to an output voltage ratio of $\frac{V_{o1}}{V_{o2}} = 3.50$. Therefore, the SNR of the two strain gauge sensors can be derived as $\frac{SNR_1}{SNR_2} = 3.50$ in theory. This means that the

Table 6.1 Main parameters of a compliant dual-stroke micropositioning stage

Parameter	Symbol	Value	Unit
Length of flexures associated with bearing #1	l_1	35	mm
In-plane width of flexures associated with bearing #1	h_1	0.35	mm
Length of flexures associated with bearing #2	l_2	20	mm
In-plane width of flexures associated with bearing #2	h_2	0.35	mm
Out-of-plane thickness of the plate material	b	10	mm
Clearance	δ	1.08	mm

Table 6.2 Stage performances evaluated by analytical models and FEA simulations

Performance	Analytical model	FEA simulation	Model error (%)
K_1 (N/m)	717.0	750.3	−4.44
K_2 (N/m)	3842.7	4006.7	−4.09
R_1^{max} (mm)	3.05	3.02	0.99
R_2^{max} (mm)	10.48	9.36	11.97
D_1 (mm)	1.28	1.268	0.95
D_2 (mm)	3.00	3.00	0
F_{max} (N)	12.30	12.82	−4.06

resolution ratio of the two strain sensors is about $\frac{1}{3.50}$; that is, the resolution in the smaller range has been improved by 3.50 times compared with that in the larger motion range.

6.4.2   FEA Simulation Results

To verify the performance of the stage, FEA simulations are conducted using the ANSYS software package.

6.4.2.1   Statics Analysis Results

The statics performance of the designed stage has been evaluated using static structural FEA. The simulations are carried out by applying an input force to produce the smaller and larger motions, respectively.

Concerning the smaller motion range, the FEA result of the deformation is illustrated in Fig. 6.5(a). To generate a motion range of ± 0.2 mm, the required driving displacement is $D_1 = \pm 1.268$ mm. Taking the FEA result as the benchmark, it is observed that the analytical model result is 0.95% higher than that of the FEA. Moreover, the simulation results reveal that the maximum value of the smaller range is $R_1^{max} = 3.02$ mm. The discrepancy between the analytical model and the FEA is about 1%. Figure 6.5(b) displays the positions at which the maximum stress occurs for bearings #1 and #2.

Regarding the larger motion range, the FEA results show that the maximum deflection is $R_2^{max} = 9.36$ mm. Compared with the FEA results, the analytical model overestimates the larger motion range by 11.97%. In addition, to generate a one-sided total motion range of 3.2 mm, the FEA predicts that the required maximum driving force is $F_{max} = 12.82$ N. The difference between the analytical and FEA results is 4.06%.

For the purpose of comparison, the results of the analytical models and the FEA simulations are tabulated in Table 6.2. The model errors mainly arise from the employed assumption for the analytical models, which only consider the bending deformations of the leaf flexures. The model accuracy can be enhanced by means of nonlinear modeling.

6.4.2.2   Dynamics Analysis Results

The dynamics performance of the stage has been evaluated by conducting the modal analysis simulation. To generate a better assessment through FEA simulation, all of the moving components including the moving coil of the VCM, the mechanical mover, and the sensor target are added to the stage. In addition, to obtain the resonant modes excited by the actuation, a cylindrical constraint is added to restrict the movement of the moving coil in the axial direction only. The simulation results of the first four resonant mode shapes are shown in Fig. 6.6. The first four resonant frequencies are: 20.58, 68.45, 120.96, and 172.90 Hz, respectively.

It is observed that the first mode shape indicates a translation along the working direction with a resonant frequency of 20.58 Hz, whereas the second mode at 68.45 Hz is attributed to the in-plane translations of the moving coil and the output platform in opposite directions. The third mode at 120.96 Hz is caused by the rotational motion of bearing #1. In addition, the fourth mode at 172.90 Hz is contributed by the translations of the intermediate flexures

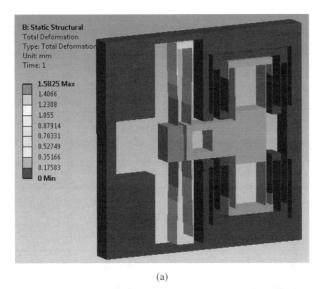

(a)

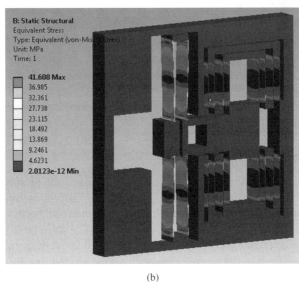

(b)

Figure 6.5   Static structural FEA simulation results: (a) deformation shape; (b) stress distribution.

associated with the two bearings. Note that the third and fourth resonant modes only induce slight movement of the output platform.

6.5   Prototype Development and Experimental Studies

In this section, a prototype micropositioning stage is described and its performance verified by means of experimental studies.

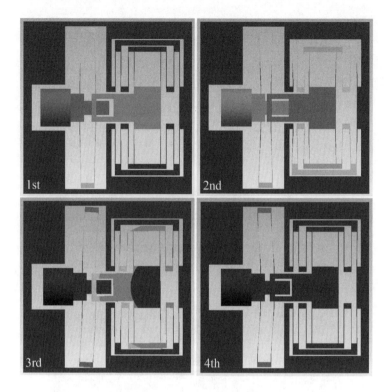

Figure 6.6   The first four resonant mode shapes of the dual-stroke stage.

6.5.1   Prototype Development

Figure 6.7 depicts a CAD model of the stage. A photograph of the prototype stage is shown in Fig. 6.8. The prototype is fabricated from a plate of Al 7075 alloy by the wire-electrical discharge machining (EDM) process. Note that the four intermediate flexures in bearing #1 of the prototype have been cut to comply with the design. The stage possesses dimensions of 110 mm × 100 mm × 10 mm. A VCM (model: NCC04-10-005-1A, from H2W Techniques, Inc.) is selected to drive the stage by considering the stroke and force requirements. It provides the maximum driving force of 29.2 N with a stroke of 10.2 mm. The VCM is driven by the NI-9263 analogy output module (from National Instruments Corp.) through a VCM driver. The stage output displacements in the smaller and larger ranges are measured by two strain gauges (model: SGD-3/350-LY13, from Omega Engineering Ltd.). The strain gauge has a nominal resistance of 350 Ω, a gauge factor of 2, and a dimension of 7 mm × 4 mm.

To measure the output voltage of the quarter-bridge circuit as shown in Fig. 6.4, the NI-9945 quarter-bridge completion accessory is used to complete the 350 Ω sensor. The NI-9945 contains three high-precision resistors of 350 Ω. Generally, the output voltage of a Wheatstone bridge is very small. In order to measure the bridge output accurately, a voltage amplifier can be adopted to pre-amplify the sensor output. However, the sensor noise will be amplified at the same time. Alternatively, the bridge output can be acquired by using a high-resolution data acquisition device directly. In this work, the NI-9237 bridge input module is employed, which

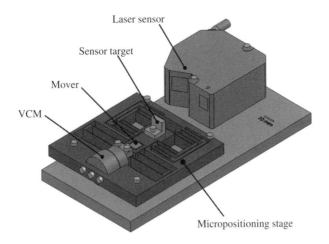

Figure 6.7 CAD model of the dual-stroke micropositioning stage.

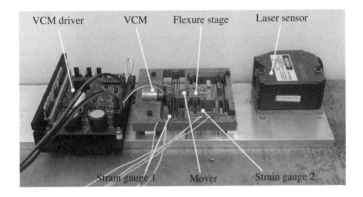

Figure 6.8 Fabricated prototype of the dual-stroke micropositioning stage.

provides 24-bit resolution. It is able to acquire the quarter-bridge output signal directly and produce a maximum voltage output of ±25 mV per volt of excitation voltage. For example, with an excitation voltage of 3.3 V, the maximum bridge output is ±82.5 mV.

For the calibration of the strain sensors, a laser displacement sensor (model: LK-H055, from Keyence Corp.) with a sub-micrometer resolution and a measurement range of 20 mm is employed. In addition, an NI cRIO-9022 real-time controller combined with an NI-9118 chassis is adopted as the control hardware. NI LabVIEW software is employed to implement deterministic real-time control of the micropositioning system.

6.5.2 Statics Performance Testing

The sensor resolutions in the two motion ranges are tested first. By applying a sinusoidal voltage signal with a frequency of 0.1 Hz and an amplitude of 0.15 V, as shown in Fig. 6.9(a), to

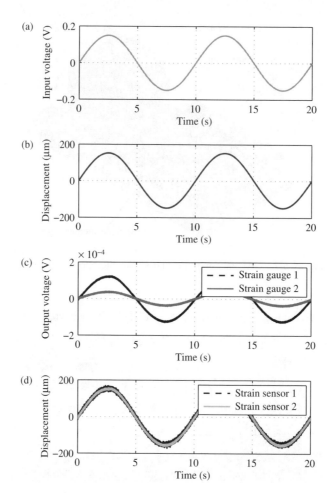

Figure 6.9 Experimental calibration results of the two strain sensors: (a) input voltage; (b) output displacement measured by the laser sensor; (c) output voltages of the two strain gauges; (d) output displacement of the two strain sensors.

the VCM driver, the output displacements of the stage measured by the laser sensor are shown in Fig. 6.9(b). In addition, the output displacements of the stage are measured by the two strain gauges and the laser displacement sensor, as shown in Fig. 6.9(c) and (d), respectively.

It is observed that the two strain gauges produce voltage ranges of 2.5485×10^{-4} and 8.2139×10^{-5} V, respectively. This leads to an output voltage ratio of $\frac{V_{o1}}{V_{o2}} = 3.10$. Assume that the noises of the two sensors exhibit the same magnitude, then the SNR of the two strain gauge sensors can be derived as 3.10, which is close to the analytically predicted value of 3.50.

The two strain-gauge sensors are calibrated by comparing their output voltages to the laser displacement sensor output. The calibrated sensors provide the output displacements as shown in Fig. 6.9(d), which indicates that the output of sensor #2 is noisier than that of sensor #1.

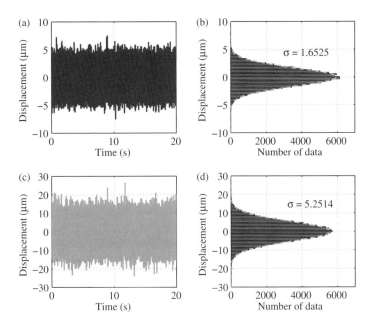

Figure 6.10 (a) Time history of the measured noises of strain sensor #1 and (b) the noise histogram; (c) time history of the measured noises of strain sensor #2 and (d) the noise histogram.

Figure 6.10 shows the noises of the two sensors, which are recorded under zero-voltage input. The histograms show that the noises follow the normal distribution closely, with standard deviations (σ) of 1.6525 and 5.2514 µm, respectively. By adopting 3σ as the resolution, the resolutions of the fine and coarse sensors are calculated as 4.9575 and 15.7542 µm, respectively. Hence, the minimum steps of this stage that can be detected in the fine and coarse ranges are 4.9575 and 15.7542 µm, respectively. In addition, the resolution ratio of the two sensors can be derived as $\frac{1}{3.18}$, which is very close to the analytical prediction of $\frac{1}{3.50}$. Therefore, the resolution in the smaller range has been improved by 3.18 times compared with that in the larger motion range.

Next, the magnitudes of the two motion ranges are tested by applying a sinusoidal signal with an amplitude of 3 V, as shown in Fig. 6.11(a). The output displacements are measured by the two strain sensors as plotted in Fig. 6.11(b) and (c), respectively. Sensor #1 saturates in the range around [−220 µm, 230 µm], which represents the smaller motion range of the stage. In the larger motion ranges of [−2800 µm, −220 µm] and [230 µm, 3200 µm], the output displacement is measured by sensor #2 alone, which provides a worse resolution than sensor #1, as tested earlier.

6.5.3 Dynamics Performance Testing

The dynamics performance of the stage is examined by the frequency response method. Specifically, a swept-sine signal with an amplitude of 0.02 V and a frequency range of 1–500 Hz is applied to drive the VCM. The frequency responses of the stage output position are shown in the Bode plots in Fig. 6.12, which are generated from the outputs of the laser sensor and

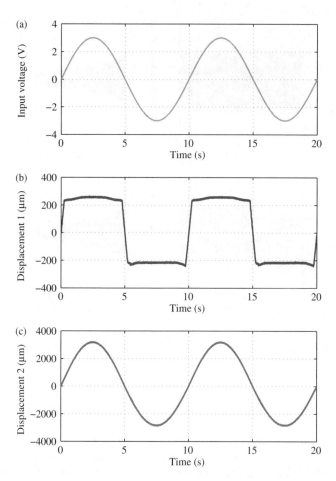

Figure 6.11 Experimental results of motion range tests: (a) input voltage; (b) smaller motion range measured by strain sensor #1; (c) larger motion range measured by strain sensor #2.

two strain sensors. The two strain-gauge sensors capture the first three modes at 16.7, 44.9, and 117.5 Hz, respectively. In contrast, the laser sensor is only able to detect the first two modes.

The reason why the laser sensor cannot capture the third resonant mode can be explained by considering the measurement principles of the sensors. Specifically, the two strain sensors measure the displacement of the output platform indirectly by monitoring the strain deformation of the internal flexures associated with the two compliant bearings. The laser sensor is fixed at the base, and it solely measures the output displacement of the platform through the sensor target, which is attached to the platform. As predicted by the FEA simulation result shown in Fig. 6.6, the first and second modes are attributed to the translations of the output platform and the two bearings. Thus, they are detected by all three sensors. However, the third resonant mode at 117.5 Hz is dominantly contributed by the deformation of the intermediate flexures, as evaluated by the FEA result shown in Fig. 6.6(c), while the output platform only

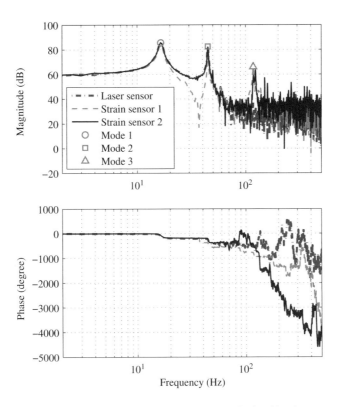

Figure 6.12   Bode plots of the frequency responses obtained by three sensors.

experiences a negligible translation. Hence, this mode is not detected by the laser sensor, which only monitors the output platform displacement, but it is captured by the two attached strain sensors. The experimental results verify the resonant mode shapes predicted by the FEA simulation.

6.5.4   Further Discussion

For a better understanding of the results, the desired and actual performances of the stage are summarized in Table 6.3. The smaller motion range is larger than the design goal of ±0.2 mm, while the larger range is slightly lower than the design specification of ±3.2 mm. The overall motion range of 6.0 mm is very close to the design objective of 6.4 mm, which also indicates that the driving displacement lies within the stroke of the employed VCM actuator. The phenomenon that the statics testing results of the bidirectional motion range are not exactly symmetric with respect to zero may be attributed to fabrication errors and the unequal clearance between the mover and each stopper. The discrepancy between the desired and actual resolution ratios for the two ranges may be caused by the manufacturing tolerance and the variations in the adhesive layers of the two strain gauges.

Table 6.3 Comparison of the desired and actual performances of the dual-stroke stage

Performance	Desired	Experimental result
Smaller range (mm)	[−0.2, 0.2]	[−0.22, 0.23]
Larger range (mm)	[−3.2, −0.2] & [0.2, 3.2]	[−2.8, −0.22] & [0.23, 3.2]
Coarse/fine resolution ratio	3.50	3.18

The two strain sensor outputs are compared under the assumption that they are glued on the flexures under the same conditions in terms of the thickness of the adhesive layers and the operating temperature, etc. The resolutions of the two strain sensors are tested using the quarter-bridge circuits without adopting filters. The resolutions can be improved two fold or four fold by using half-bridge or full-bridge circuits, respectively. Moreover, the resolution can be further enhanced by removing the low-frequency noises of the sensor outputs.

Regarding the dynamics performance testing, all three sensors predict a first-mode frequency of 16.7 Hz, which is 18.85% lower than the FEA simulation result of 20.58 Hz. The discrepancy between the simulation and experimental results for the resonant frequencies mainly comes from the fabrication error in the stage parameters. In practice, input signals with frequency component around 17 Hz will excite the first resonant mode, and the first natural frequency limits the usable bandwidth of the system. Thus, a high first natural frequency is important to achieve a large control bandwidth.

The drawback of the proposed stage is that the fine resolution only applies to a smaller positioning range, although the coarse resolution can be used in both smaller and larger ranges. Hence, fine tuning can only be implemented in the smaller motion range around the home position of the stage. When the lower stiffness mechanism impacts the stopper, the flexure may vibrate and the larger motion may be affected by the impact. The impact effect cannot be observed in Fig. 6.11(b) and (c) because these results are generated by applying an input signal with a 0.1-Hz low frequency. An input signal with a higher frequency (e.g., 5 Hz) will lead to an apparent impact effect. Yet, unlike the interference in dual-servo stages, where the interaction occurs during the entire positioning of the fine stage, the impact effect in the presented dual-range stage only appears at the two limits of the smaller motion range. To avoid this impact, a low positioning speed at the two limits can be commanded.

The reported dual-range stage has promising applications in precision positioning situations. For instance, the dual-range stage can be employed to execute a microassembly task, as illustrated in Fig. 6.13. In such a case, the dual-range stage is adopted as an X stage to carry a microgripper. More details about the gripper design can be found in [10]. The fragile micro-objects to be assembled are supported by a coarse YZ stage. First, the object is transported to the vicinity of the gripper tips. If the X stage only provides a coarse resolution [11], an accurate alignment of the gripper tips relative to the object may not be realized in the x-axis direction. As a result, the fragile object is prone to be damaged or not grasped by the gripper. Alternatively, by employing a dual-range stage as the X stage, the x-axis position of the gripper tips can be finely adjusted using the smaller range of the stage for precise gripping. Once the object is grasped, the larger range of the X stage is used to transport the object to a planned assembly destination.

It is notable that the resolutions of the stage have been characterized under open-loop drive status. In future work, closed-loop control will be realized to determine the positioning

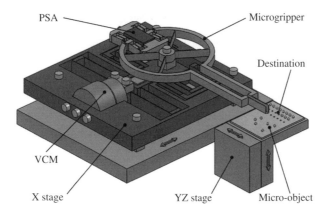

Figure 6.13 Illustration of microassembly application using the proposed dual-stroke X stage.

accuracy and repeatability of the stage. Moreover, the presented ideas can also be extended to the design of multi-axis micropositioning stages.

6.6 Conclusion

The design and verification of a uniaxial compliant micropositioning stage with dual ranges and dual resolutions was presented in this chapter. Based on the concept of unequal stiffness, a proof-of-concept design of a flexure-based stage is proposed. Analytical models have been established to predict the motion ranges, coarse/fine resolution ratio, and driving force and stroke, all of which are verified by finite-element analysis and experimental studies. The results show that the single VCM is able to produce dual motion ranges and the same kind of strain gauge sensor is capable of providing fine and coarse resolutions in the smaller and larger ranges, respectively. In future work, control schemes will be developed to accomplish a precise positioning for related applications.

References

[1] Song, Y., Wang, J., Yang, K., Yin, W., and Zhu, Y. (2010) A dual-stage control system for high-speed, ultra-precise linear motion. *Int. J. Adv. Manuf. Technol.*, **48**, 633–643.

[2] Suh, S.M., Chung, C.C., and Lee, S.H. (2002) Design and analysis of dual-stage servo system for high track density HDDs. *Microsyst. Technol.*, **8** (2&3), 161–168.

[3] Al Mamun, A., Mareels, I., Lee, T.H., and Tay, A. (2003) Dual stage actuator control in hard disk – a review, in *Proc. 29th Annual IEEE Industrial Electronics Society Conf.*, pp. 2132–2137.

[4] Xu, Q. (2012) Design and development of a flexure-based dual-stage nanopositioning system with minimum interference behavior. *IEEE Trans. Automat. Sci. Eng.*, **9** (3), 554–563.

[5] Albu-Schaffer, A., Wolf, S., Eiberger, O., Haddadin, S., Petit, F., and Chalon, M. (2010) Dynamic modelling and control of variable stiffness actuators, in *Proc. IEEE Conf. on Robotics and Automation*, pp. 2155–2162.

[6] Palli, G., Berselli, G., Melchiorri, C., and Vassura, G. (2011) Design of a variable stiffness actuator based on flexures. *J. Mech. Robot.*, **3** (3), 034 501.

[7] Fleming, A.J. and Leang, K.K. (2010) Integrated strain and force feedback for high-performance control of piezo-electric actuators. *Sens. Actuator A-Phys.*, **161** (1&2), 256–265.

[8] Xu, Q. (2012) New flexure parallel-kinematic micropositioning system with large workspace. *IEEE Trans. Robot.*, **28** (2), 478–491.

[9] Wheeler, A.J. and Ganji, A.R. (2009) *Introduction to Engineering Experimentation*, Prentice Hall, Englewood Cliffs, NJ, 3rd edn.

[10] Xu, Q. (2014) Design and smooth position/force switching control of a miniature gripper for automated micro-handling. *IEEE Trans. Ind. Informat.*, **10** (2), 1023–1032.

[11] Xu, Q. (2013) Design, testing and precision control of a novel long-stroke flexure micropositioning system. *Mech. Mach. Theory*, **70**, 209–224.

7

Multi-Stroke, Multi-Resolution XY Flexure Stage

Abstract: This chapter presents the conceptual design of a multi-stroke, multi-resolution micropositioning stage driven by a single actuator for each working axis. It eliminates the issue of interference among different drives, which occurs in conventional multi-actuation stages. The stage is devised based on a fully compliant variable-stiffness mechanism, which exhibits unequal stiffnesses in different strokes. Resistive strain sensors are employed to offer variable displacement resolutions in the different strokes. Analytical models are established to facilitate the design of the motion strokes and coarse/fine resolution ratio, verified through finite-element analysis simulations. A proof-of-concept prototype XY stage is designed, fabricated, and experimentally tested to demonstrate the feasibility of the presented ideas.

Keywords: Two-dimensional micropositioning, XY flexure mechanisms, Compliant mechanisms, Translational stages, Multi-stroke stages, Multi-range design, Multi-resolution design, Finite-element analysis, Strain-gauge sensors, Variable-stiffness mechanism.

7.1 Introduction

Multi-stroke stages are demanded in micro-/nanopositioning applications which require smaller and larger motion strokes with fine and coarse resolutions, respectively [1]. Here, the concept as reported in Chapter 6 is extended to devise a multi-stroke, multi-resolution micropositioning stage using a single drive for each axis.

The essential principle of the stage design is to devise a fully compliant variable-stiffness mechanism by employing stiff and compliant flexure bearings which are connected in series. The deformations of these flexure bearings are sequentially limited by stroke stoppers and unequal stiffnesses in different strokes are obtained. The larger and smaller deflections of the flexures are detected by using resistive sensors to offer fine and coarse displacement resolutions for different strokes. Unlike multi-servo stages, the presented technique enables the achievement of multi-stroke motion by adopting a single actuator in a working direction. This allows the elimination of the conventional interference issue. Moreover, the single-drive design permits a reduction in hardware costs and control design workload.

Design and Implementation of Large-Range Compliant Micropositioning Systems, First Edition. Qingsong Xu.
© 2016 John Wiley & Sons Singapore Pte Ltd. Published 2016 by John Wiley & Sons Singapore Pte Ltd.

Based on a uniaxial multi-stroke stage design, a two-axis micropositioning stage is devised for illustration in this chapter. Both simulation and experimental investigations have been carried out to verify the proof-of-concept design of a multi-stroke, multi-resolution XY micropositioning system.

7.2   Conceptual Design

The conceptual design of a micropositioning stage with multiple strokes and multiple resolutions is presented in the following discussion.

7.2.1   Design of Flexure Stage with Multiple Strokes

Figure. 7.1(a) shows the principle of a single-axis, multi-stroke micropositioning stage. The $n + 1$ masses (M_1 to M_{n+1}) are connected in series through n flexure bearings. The ith bearing exhibits an equivalent stiffness K_i ($i = 1$ to n). The first mass M_1 is driven by a linear actuator, and the last mass M_{n+1} is fixed at the base. To generate multiple strokes for the output platform M_n, n modules of mover and stroke stopper are connected to the n masses as shown in Fig. 7.1(a). Concerning the ith module, the mover is linked to the mass M_i, and the corresponding stroke stopper is attached to the next mass M_{i+1}. A clearance δ_i between the ith mover and each side of the ith stopper permits a restricted bidirectional translation of the mover and the mass M_i relative to the mass M_{i+1}.

Assume that the relationship $K_1 < K_2 < \cdots < K_n$ holds. Referring to Fig. 7.1(a), if the actuator drives M_1 to move forward, then all of the bearings experience a common compressive force.

Initially, no mechanical mover makes contact with the corresponding stroke stopper, and the driving force F_{s1} is experienced by all of the flexure bearings. Considering the serial connections of the equivalent linear springs, the equivalent stiffness K_{s1} of the system in the stroke

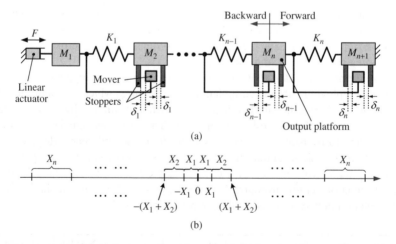

(a)

(b)

Figure 7.1   Schematic of (a) a micropositioning stage with multiple strokes and (b) the motion ranges

$[0, X_1]$ can be derived as

$$K_{s1} = \frac{1}{\sum_{j=1}^{n} \frac{1}{K_j}}. \tag{7.1}$$

When a certain driving displacement D_1 is undergone, the first mechanical mover translates over a distance δ_1 relative to M_2, and it contacts the forward stroke stopper. The corresponding displacement X_1 of the output platform M_n can be calculated by considering the following relationship:

$$F_{s1} = K_{s1}D_1 = K_n X_1 \tag{7.2}$$

which gives

$$X_1 = \frac{K_{s1}}{K_n}D_1. \tag{7.3}$$

Afterwards, if driving continues in the forward direction, the first flexure bearing K_1 will not be deformed any more because it has already been restricted by the stroke stopper attached to M_2. Thus, M_1 can be considered to be linked to M_2 by K_1 through a rigid connection. As a consequence, M_1, K_1, and M_2 are combined as a rigid body, translating in the forward direction together. Under this scenario, the overall stiffness of the mechanism becomes

$$K_{s2} = \frac{1}{\sum_{j=2}^{n} \frac{1}{K_j}}. \tag{7.4}$$

After the first mover comes into contact with the first stroke stopper, an additional driving displacement D_2 is produced to make M_2 translate a distance δ_2 relative to M_3; that is, the second mover also contacts its stroke stopper. The produced displacement X_2 of the output platform M_n can be calculated from the relationship

$$F_{s2} = K_{s2}D_2 = K_n X_2 \tag{7.5}$$

which gives

$$X_2 = \frac{K_{s2}}{K_n}D_2. \tag{7.6}$$

Similarly, after the $(i-1)$th mover contacts the corresponding stroke stopper, the overall equivalent stiffness of the system can be modeled as

$$K_{si} = \frac{1}{\sum_{j=i}^{n} \frac{1}{K_j}}. \tag{7.7}$$

The corresponding additional driving force is calculated as

$$F_{si} = K_{si}D_i = K_n X_i \tag{7.8}$$

which leads to an incremental value of the output displacement:

$$X_i = \frac{K_{si}}{K_n}D_i. \tag{7.9}$$

Hence, the forward motion range of M_n is divided into n sequential strokes $[0, X_1]$, $[X_1, X_1 + X_2]$, $\cdots$, $[X_1 + X_2 + \cdots + X_{n-1}, X_1 + X_2 + \cdots + X_n]$ as shown in Fig. 7.1(b).

Additionally, the backward motion range of the output platform is also divided into n intervals by the stroke stoppers when it is driven to translate in the backward direction.

The overall motion range is partitioned into n sequential portions $[0, |X_1|]$, $[|X_1|, |X_1 + X_2|]$, $\cdots$, $[|X_1 + X_2 + \cdots + X_{n-1}|, |X_1 + X_2 + \cdots + X_n|]$, respectively. The overall stiffness of the system exhibits unequal value in the n strokes; that is, the system stiffness has n discrete values in the entire motion range.

It is notable that the mass M_n is selected as the output platform because the associated mechanical mover is the last one that will be restricted by its stroke stopper. In contrast, if the mover attached to M_1 is the last restricted one, M_1 can be chosen as the output platform.

In the following, a multi-resolution stage is devised based on the aforementioned multi-stroke conceptual design.

7.2.2   Design of Flexure Stage with Multiple Resolutions

Recalling the foregoing analysis, it is observed that in the ith stroke $[|X_1 + X_2 + \cdots + X_{i-1}|, |X_1 + X_2 + \cdots + X_i|]$, the deformations of $i-1$ flexure bearings have already been restricted by the stroke stoppers, and the output displacement is contributed by the deformations of the remaining $n - i + 1$ bearings.

In the first stroke $[0, |X_1|]$, the output displacement is contributed by all of the bearings. Let Δ_i be the deformation that is experienced by the ith flexure bearing K_i. These deformations can be related by

$$K_1 \Delta_1 = K_2 \Delta_2 = \cdots = K_n \Delta_n \tag{7.10}$$

which describes the driving force of the actuator.

Assume that the bearings are designed to satisfy the condition

$$K_1 < K_2 < \cdots < K_n. \tag{7.11}$$

Then, it can be deduced from Eqs. (7.10) and (7.11) that

$$\Delta_1 > \Delta_2 > \cdots > \Delta_n \tag{7.12}$$

which indicates that the deformation of bearing i is greater than that of the adjacent next bearing $i + 1$.

It is known that the resistive type of sensors (e.g., resistive strain gauges [2] or piezoresistive sensors [3]) can be used to measure the displacement of a flexure mechanism indirectly by detecting the induced resistance changes. If the same resistive sensor is adopted to measure two deformations with different magnitudes, the larger the deformation is, the larger the output signal will be. That is, a larger deformation results in a higher signal-to-noise ratio (SNR), i.e., a higher resolution for the displacement measurement. By mounting n resistive sensors on the n flexure bearings, the displacement of the output platform can be measured by these sensors. The relationship of the resolutions provided by these sensors can be derived below in view of Eq. (7.12):

$$\text{Resol}_1 > \text{Resol}_2 > \cdots > \text{Resol}_n. \tag{7.13}$$

In the first stroke, the first resistive sensor, which is attached on the first flexure bearing, provides the best resolution for the output displacement measurement. Meanwhile, in the second stroke, the deformation of the first bearing remains unchanged, and hence the first sensor is saturated since its reading does not change. So, the second sensor, which is mounted on the second bearing, provides the best resolution. For the last stroke interval n, the nth sensor delivers a worst resolution while the other sensors are all saturated. In most practical applications, a fine resolution is needed in a smaller stroke, and a coarser resolution is acceptable for a larger stroke. Hence, it is meaningful to design the multiple strokes as follows:

$$X_1 < X_2 < \cdots < X_n. \tag{7.14}$$

In this way, a micropositioning stage with multiple strokes as well as multiple resolutions is devised. Specifically, the higher and lower resolutions are generated in the smaller and larger motion strokes, respectively.

7.3 Flexure-Based Compliant Mechanism Design

To illustrate the presented design idea, the mechanism design of flexure-based micropositioning stages with dual strokes and resolutions is proposed in this section. Note that the design can easily be extended to obtain more than two strokes and resolutions.

7.3.1 Compliant Element Selection

In the literature, various types of flexible elements are available for a compliant mechanism design. To devise a micropositioning stage with long stroke, the leaf flexures are employed in this work. In particular, two basic flexure elements as shown in Fig. 7.2(a) and (b) are adopted to design the bearings. Element #1 consists of two fixed–guided flexures, which experience an identical magnitude of deformation. Element #2 employs a MCPF with two modules ($N = 2$). Element #2 involves eight fixed–guided leaf flexures with the same dimensions. Applying a force F at the output end, elements #1 and #2 permit large output displacements of Δ_1 and Δ_2, respectively.

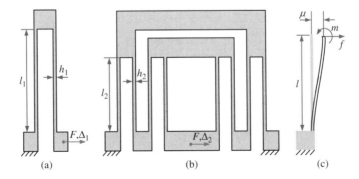

Figure 7.2   Basic flexure element #1 (a) and element #2 (b); (c) deflection of a fixed–guided leaf flexure

Based on these two elements, a uniaxial, dual-stroke, dual-resolution flexure stage is designed in Chapter 6. As an extension, the mechanism design of a two-axis, dual-stroke, dual-resolution flexure stage is presented in the next section.

7.3.2 Design of a Two-Axis Stage

For illustration, an embodiment of a two-axis XY micropositioning stage with dual strokes and resolutions is devised as shown in Fig. 7.3. Owing to its symmetric design, the translations in the x and y directions follow the same working principle.

The x-axis motion is produced by the bearings $x1$ and $x2$, which are composed of two elements #1 and six elements #2, respectively. The mover $x1$ is connected to the driving end of actuator #1, and the stroke stopper $x2$ is fixed at the base. In addition, the stopper $x1$ and mover $x2$ are formed by the holes in the output platform. The bidirectional translations of the movers $x1$ and $x2$ are restricted by the stoppers $x1$ and $x2$, respectively. The x-axis motion is guided by the bearings $x1$ and $x2$. Similarly, the y-axis motion is guided by the bearings $y1$ and $y2$ accordingly.

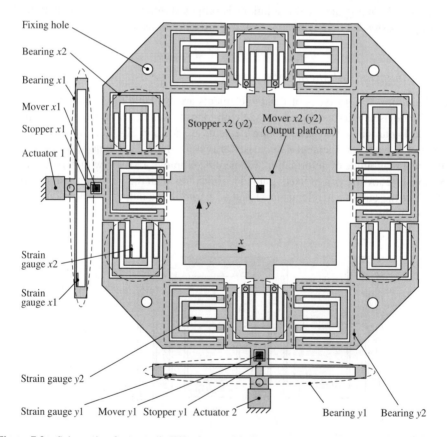

Figure 7.3 Schematic of a two-axis XY micropositioning stage with dual strokes and resolutions

It is notable that the bearings $x2$ and $y2$ construct a decoupled XY flexure stage, which guarantees that the translational motion in the x (y) direction is independent of that in the y (x) direction. Compared with the XY stage proposed in previous work [4], the presented XY stage has a modified fixing scheme, which is the same as the one illustrated in Fig. 3.3(e). Moreover, the XY stage has two extra bearings $x1$ and $x2$ to accomplish the objective of multi-stroke and multi-resolution design. To achieve a specified performance, the stage parameters are designed in the following section.

7.4 Parametric Design

To produce a stage with the desired performance in terms of motion ranges and resolutions, the flexure parameters demand a quantitative design. Taking the XY micropositioning stage as an example, the parametric design of the stage for the specified motion strokes and coarse/fine sensor resolution ratio is presented in this section.

Due to its symmetric structure, the following design is conducted by considering the x-axis motion only. The same parametric design applies to the y-axis motion.

7.4.1 Design of Motion Strokes

7.4.1.1 Stiffness Modeling of Basic Elements

First, the equivalent stiffnesses of the two basic elements as shown in Fig. 7.2(a) and (b) are modeled.

The leaf flexures in each of the two elements can be considered as fixed–guided beams, as shown in Fig. 7.2(c). Recalling the modeling of a fixed–guided beam as shown in Eqs. (2.1) and (2.2), the linear stiffness of one leaf flexure is derived as

$$K_0 = \frac{Ebh^3}{l^3} \tag{7.15}$$

where E is the Young's modulus of the material.

Referring to Fig. 7.2(a), taking into account that element #1 is composed of two leaf flexures which are connected in series, its equivalent stiffness can be calculated as follows:

$$K_{e1} = \frac{K_0}{2} = \frac{Ebh_1^3}{2l_1^3} \tag{7.16}$$

where l_1 and h_1 represent the in-plane length and width of the leaf flexures in element #1, respectively.

In addition, taking into account that element #2 shown in Fig. 7.2(b) belongs to a MCPF with $N = 2$, its stiffness can be derived below in view of Eq. (2.11):

$$K_{e2} = \frac{Ebh_2^3}{2l_2^3} \tag{7.17}$$

where l_2 and h_2 represent the length and width of the leaf flexures associated with element #2, respectively.

7.4.1.2 Stress and Deflection Calculation

For each leaf flexure, the bending deflection dominates the deformation. Hence, the minor axial deformation is neglected in the analytical modeling process.

Taking into account the bending deformation of each leaf flexure alone, the maximum allowable stress σ_{max} (i.e., the yield strength σ_y) occurs around the two terminals when the flexure undergoes the maximum moment m_{max}. The stress is calculated as

$$\sigma_{max} = \frac{m_{max}h}{2I} = \sigma_y \tag{7.18}$$

which gives the maximum moment

$$m_{max} = \frac{\sigma_y bh^2}{6} \tag{7.19}$$

where m_{max} is induced by the applied force f_{max}.

Similar to Eq. (2.9), the maximum deflection of one leaf flexure is generated as

$$\mu_{max} = \frac{\sigma_y l^2}{3Eh}. \tag{7.20}$$

Additionally, it is deduced that the deflections of the leaf flexures in bearings $x1$ and $x2$ are equal to one-half and one-quarter of the deformations of bearings $x1$ and $x2$, respectively. That is,

$$\mu_1 = \frac{\Delta_1}{2} \tag{7.21}$$

$$\mu_2 = \frac{\Delta_2}{4}. \tag{7.22}$$

7.4.1.3 Stiffness Modeling of the XY Stage

Recalling that bearings $x1$ and $x2$ are the parallel connections of two elements #1 and six elements #2, respectively, their stiffnesses can be derived in view of Eqs. (7.16) and (7.17):

$$K_1 = 2K_{e1} = \frac{Ebh_1^3}{l_1^3} \tag{7.23}$$

$$K_2 = 6K_{e2} = \frac{3Ebh_2^3}{l_2^3}. \tag{7.24}$$

In consideration of Eqs. (7.1) and (7.4) along with $n = 2$, the equivalent stiffnesses of the XY stage in the two strokes can be expressed as

$$K_{s1} = \frac{1}{\frac{1}{K_1} + \frac{1}{K_2}} \tag{7.25}$$

$$K_{s2} = K_2. \tag{7.26}$$

7.4.1.4   Smaller Stroke Design

Assume that the two motion ranges are $[0, |X_1|]$ and $[|X_1|, |X_1 + X_2|]$, respectively. The absolute deformation of bearing $x1$ is derived as follows:

$$\delta_1 = D_1 - X_1 \tag{7.27}$$

where D_1 is the driving displacement of the actuator.

Taking into account Eqs. (7.3) and (7.25), the required driving displacement from the actuator can be obtained as

$$D_1 = \left(1 + \frac{K_2}{K_1}\right) X_1. \tag{7.28}$$

Then, substituting Eq. (7.28) into Eq. (7.27), a fundamental algebra operation gives

$$X_1 = \frac{K_1}{K_2}\delta_1. \tag{7.29}$$

Considering Eqs. (7.23) and (7.24), the above Eq. (7.29) can be expressed in the form

$$X_1 = \frac{h_1^3 l_2^3}{3h_2^3 l_1^3}\delta_1 \tag{7.30}$$

which indicates that the smaller motion range X_1 is governed by the clearance δ_1 as well as the parameters l_1, h_1, l_2, and h_2 of the two flexure bearings.

Meanwhile, to avoid plastic deformation of the flexures, the stress caused by the deflection Δ_1 of bearing $x1$ should not exceed the yield strength of the material. Considering Eqs. (7.20) and (7.21), the allowable maximum deformation of bearing $x1$ can be computed as

$$\Delta_1^{\text{allow}} = \frac{2\sigma_y l_1^2}{3Eh_1} \tag{7.31}$$

where σ_y denotes the yield strength of the material.

In practice, Δ_1 is constrained by the clearance δ_1 between the mover $x1$ and the stroke stopper $x1$. Hence, this clearance should be designed to meet the following relationship:

$$\delta_1 \leq \Delta_1^{\text{allow}}. \tag{7.32}$$

Taking into account Eqs. (7.30) and (7.32), the upper limit of the one-sided smaller stroke is determined as

$$X_1^{\text{max}} = \frac{2\sigma_y h_1^2 l_2^3}{9Eh_2^3 l_1}. \tag{7.33}$$

7.4.1.5   Larger Stroke Design

The larger motion stroke X_2 is produced by the deformation of bearing $x2$ solely, because the deformation of bearing $x1$ is stopped by the stopper $x1$. In view of Eq. (7.6) with $n = 2$ and Eq. (7.26), the required driving displacement from the actuator is determined as

$$D_2 = X_2. \tag{7.34}$$

Note that bearing $x2$ is deformed in both smaller and larger strokes, and the overall deformation of bearing $x2$ is constrained by the clearance δ_2 between the mover $x2$ and the stroke stopper $x2$. That is, the overall motion range of the stage is limited by δ_2. Hence, the one-sided larger stroke can be computed as

$$X_2 = \delta_2 - X_1. \tag{7.35}$$

Considering Eqs. (7.20) and (7.22) together, the allowable maximum deformation of bearing $x2$ can be calculated as

$$\Delta_2^{\text{allow}} = \frac{4\sigma_y l_2^2}{3Eh_2}. \tag{7.36}$$

Practically, Δ_2 is constrained by the clearance δ_2. Hence, to avoid plastic deformation of the flexures associated with bearing $x2$, the clearance δ_2 should be designed to meet the condition

$$\delta_2 \leq \Delta_2^{\text{allow}}. \tag{7.37}$$

In view of Eqs. (7.35) and (7.37), once the smaller stroke X_1 is specified, the upper limit of the one-sided larger stroke is determined below:

$$X_2^{\text{max}} = \frac{4\sigma_y l_2^2}{3Eh_2} - X_1. \tag{7.38}$$

7.4.2 Design of Coarse/Fine Sensor Resolution Ratio

In this work, two sets of resistive strain gauges are employed to measure the displacement of the compliant stage in the smaller and larger motion strokes, respectively. To enhance the sensor sensitivity, the strain gauges are mounted around the places of largest stress of the leaf flexures associated with bearings $x1$ and $x2$, respectively.

Assume that the output displacement in each axis of the XY stage is measured by the two strain-gauge sensors through two Wheatstone bridge circuits as shown in Fig. 6.4. The bridge output voltages can be calculated approximately as follows [5]:

$$V_o \approx \frac{kV_s}{4R} \times dR \tag{7.39}$$

where $k = 1, 2,$ and 4 represent the situations of quarter-, half-, and full-bridge circuits, respectively. Besides, V_s is the supply source voltage. dR and R represent the change and nominal values of the gauge resistance, respectively.

The gauge factor of a strain gauge is expressed as

$$S = \frac{dR/R}{\epsilon} \tag{7.40}$$

where ϵ is the strain induced by the deformation of the flexure bearing. The strain ϵ is related to the experienced stress σ of the flexure by

$$\sigma = E\,\epsilon \tag{7.41}$$

where E is the Young's modulus of the material.

Considering Eq. (7.20) along with σ_y replaced by σ, a relationship between the guided deflection μ and stress σ is obtained:

$$\mu = \frac{\sigma l^2}{3Eh}. \tag{7.42}$$

In view of Eqs. (7.39)–(7.42), the relationship between the circuit output voltage and the flexure deflection can be derived:

$$\mu = \frac{4l^2 V_o}{3hSV_s}. \tag{7.43}$$

Recalling Eqs. (7.21) and (7.22), the ratio of output voltages (V_{o1} and V_{o2}) of the two strain gauges associated with the two bearings ($x1$ and $x2$) is obtained from Eq. (7.43) as follows:

$$\eta = \frac{V_{o1}}{V_{o2}} = \frac{2h_1 l_2^2 \Delta_1}{h_2 l_1^2 \Delta_2}. \tag{7.44}$$

Taking into account Eq. (7.10) as well as Eqs. (7.23) and (7.24), the relation (7.44) further reduces to

$$\eta = \frac{V_{o1}}{V_{o2}} = \frac{6h_2^2 l_1}{h_1^2 l_2} \tag{7.45}$$

which indicates that the output voltage ratio of the two strain gauges is determined by the flexure parameters l_1, h_1, l_2, and h_2 of the XY stage. By properly selecting these flexure parameters, a relation $\eta > 1$ can be obtained. That is, a higher SNR will be produced by the strain gauge $x1$. It follows that a coarse/fine resolution ratio η can be achieved by the two strain-gauge sensors.

7.4.3 Actuation Issue Consideration

A voice coil motor (VCM) is adopted to produce sufficient driving displacement. To create a total motion range of $2(X_1 + X_2)$, a driving displacement of $D_{input} = 2(D_1 + D_2)$ is needed, which can be calculated using Eqs. (7.28) and (7.34).

This input displacement should not exceed the stroke $D_{actuator}$ of the selected actuator, i.e.,

$$D_{input} = 2\left(1 + \frac{3h_2^3 l_1^2}{h_1^3 l_2^3}\right) X_1 + 2X_2 \leq D_{actuator} \tag{7.46}$$

which is calculated in consideration of Eqs. (7.23) and (7.24).

In addition, the stage should be designed to be sufficiently compliant so that the elastic energy can be overcome by the VCM. Taking into account the assumption in Eq. (7.11), it is deduced from Eqs. (7.25) and (7.26) that $K_{s1} < K_{s2}$. That is, the stiffness in the smaller stroke is lower than that in the larger one. Hence maximum force is needed to produce the extremum of the overall motion range. The required maximum driving force can be calculated as follows:

$$F_{max} = K_{s2}(X_1 + X_2) \leq F_{actuator} \tag{7.47}$$

where $F_{actuator}$ denotes the maximum driving force of the VCM actuator.

Substituting Eqs. (7.26) and (7.24) into Eq. (7.47), one can derive that

$$F_{\max} = \frac{3Ebh_2^3}{l_2^3}(X_1 + X_2) \le F_{\text{actuator}}. \tag{7.48}$$

Hence, Eqs. (7.46) and (7.48) provide guidelines for the stage parameter design in consideration of the stroke and force limits of the actuator.

7.5 Stage Performance Assessment

The foregoing parametric design procedures offer guidelines in Eqs. (7.32), (7.33), (7.37), (7.38), (7.45), (7.46), and (7.48) for the stage parameter design. As a case study, an XY micropositioning stage is designed to produce the smaller and larger motion strokes of $X_1 = Y_1 = 0.2$ mm and $X_2 = Y_2 = 2.3$ mm, respectively, as well as a coarse/fine resolution ratio of $\eta = 5.6$ in each working axis. The selected VCM provides a stroke of 10.2 mm and maximum driving force of 29.2 N. The main parameters of the stage are shown in Table 7.1.

7.5.1 Analytical Model Evaluation Results

The analytical models predict that the stage parameters allow the generation of $X_1^{\max} = 2.10$ mm, $X_2^{\max} = 11.79$ mm, $D_{\text{input}} = 8.02$ mm, and $F_{\max} = 25.0$ N, which all satisfy the aforementioned design criteria. The overall motion range in each axis is constrained by δ_2 as ±2.5 mm.

In addition, the parametric design leads to an output voltage ratio $\eta = \frac{V_{o1}}{V_{o2}} = 5.6$. Therefore, the SNR relationship of the two strain gauges can be predicted as $\frac{\text{SNR}_1}{\text{SNR}_2} = \eta = 5.6$. This means that the coarse/fine resolution ratio of the two strain gauge sensors is $\frac{\text{Resol}_2}{\text{Resol}_1} = \eta = 5.6$. That is, the resolution in the smaller stroke has been improved 5.6 times compared with that in the larger motion stroke of each working axis.

7.5.2 FEA Simulation Results

To verify the accuracy of the established analytical models, finite-element analysis (FEA) simulations are carried out using the ANSYS software package.

Table 7.1 Main parameters of a dual-stroke and dual-resolution XY micropositioning stage

Parameter	Symbol	Value	Unit
Flexure length of element #1	l_1	40	mm
Flexure width of element #1	h_1	0.5	mm
Flexure length of element #2	l_2	21	mm
Flexure width of element #2	h_2	0.35	mm
Out-of-plane thickness of plate material	b	10	mm
Clearance #1	δ_1	1.5	mm
Clearance #2	δ_2	2.5	mm

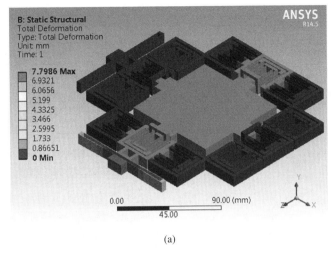

(a)

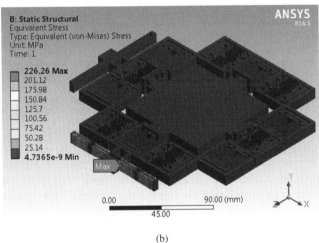

(b)

Figure 7.4 Static FEA simulation results with an input force applied: (a) deformation results; (b) stress distribution results

7.5.2.1 Statics Analysis Results

The statics performance of the XY stage is evaluated by static structural analysis with FEA. The simulations are carried out by applying an input force at the input end to produce the smaller and larger motions, respectively. Owing to the symmetric structure of the XY flexure stage, only the x-axis results are presented here.

Specifically, with an input force of 10 N applied at the input end, the performance in the smaller stroke is tested. The FEA simulation results of the deformation and stress distribution are illustrated in Fig. 7.4(a) and (b), respectively. To generate a motion stroke of ±0.2 mm, the required driving displacement is ±1.80 mm. Considering the FEA result as the benchmark, it is observed that the analytical model (±1.71 mm) underestimates the input displacement by

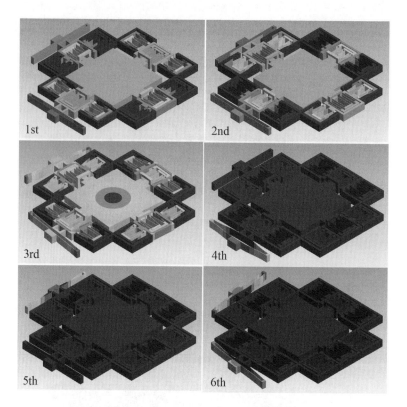

Figure 7.5   Simulation results of the first six resonant mode shapes of the dual-stroke XY stage

5%. Moreover, simulation results reveal that the maximum value of the smaller range is $X_1^{\max}$ = 2.02 mm. The discrepancy between the analytical model and FEA simulation is less than 0.5%. In addition, Fig. 7.4(b) confirms that the maximum stresses occur around the ends of the leaf flexures associated with bearings $x1$ and $x2$, respectively.

Regarding the larger motion stroke, FEA results indicate that the maximum deflection $X_2^{\max}$ is 12.10 mm. Compared with the FEA prediction, the analytical model underestimates the larger stroke by 2.6%. In addition, to generate a one-sided overall motion range of 2.5 mm, the FEA predicts that the required maximum driving force is 27.5 N. Hence, the difference between the analytical and FEA results for the driving force assessment is 9.1%.

The aforementioned discrepancies between the analytical model and FEA simulation results are all lower than 10%. These discrepancies arise mainly from the adopted assumption for the analytical models, which only considers the bending deformations of the leaf flexures. The model accuracy can be enhanced by taking into account both bending and axial deformations of the flexures.

7.5.2.2   Dynamics Analysis Results

The modal analysis is performed to evaluate the dynamics performance of the XY stage. In particular, the shapes of the first six resonant modes are obtained as shown in Fig. 7.5. Table 7.2 shows the corresponding frequencies of the first six resonant modes.

Table 7.2 First six resonant mode frequencies of the dual-stroke XY stage

Mode no.	Frequency (Hz)
1	35.17
2	35.26
3	72.51
4	94.29
5	94.41
6	115.62

It is observed that the first two mode shapes indicate the translations along the x- and y-axes with similar resonant frequencies of 35.17 and 35.26 Hz, respectively. The third mode is attributed to the in-plane rotation of the stage output platform. It is found that the third resonant frequency is 72.51 Hz, which is more than twice as high as the first two fundamental frequencies. The fourth to sixth resonant modes are induced by the translations of the stage along the two working axes at higher frequencies of 94.29, 94.41, and 115.62 Hz, respectively. Moreover, the simulation results reveal that the remaining resonant modes are mainly induced by the in-plane and out-of-plane deflections of the internal leaf flexures associated with the two bearings.

7.6   Prototype Development and Experimental Studies

In this section, a prototype XY micropositioning stage is described and the experimental testing results are presented.

7.6.1   Prototype Development

Figure. 7.6 shows a photograph of the developed prototype XY stage. The stage is fabricated from a plate of Al 7075 alloy by the wire-electrical discharge machining (EDM) process. The stage possesses dimensions of 240 mm × 240 mm × 10 mm. For actuation, two VCMs (model: NCC04-10-005-1A, from H2W Techniques, Inc.) are chosen to provide a sufficiently large stroke of 10.2 mm and output force of 29.2 N. Each VCM is driven by the NI-9263 analogy output module (from National Instruments Corp.) through a VCM driver. The stage output displacements in the smaller and larger strokes in each working axis are measured by two resistive strain gauges (model: SGD-3/350-LY13, from Omega Engineering Ltd.). This type of strain gauge possesses a nominal resistance of 350 Ω, a gauge factor of 2, and dimensions of 7 mm × 4 mm.

To measure the output voltage of the quarter-bridge circuit as shown in Fig. 6.4, the NI-9945 quarter-bridge completion accessory is used to complete the 350-Ω sensor. The NI-9945 contains three high-precision resistors of 350 Ω. The bridge output voltage is acquired by using a high-resolution data acquisition device. In this work, the NI-9237 bridge input module is employed, which provides 24-bit a resolution.

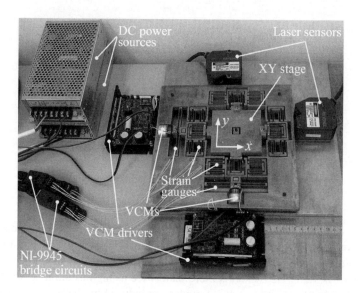

Figure 7.6   Fabricated prototype of the dual-stroke XY micropositioning stage

For the calibration of the strain sensors, two laser displacement sensors (model: LK-H055, from Keyence Corp.) with a measurement range of 20 mm are employed. In addition, an NI cRIO-9022 real-time controller combined with NI-9118 chassis is adopted as the controller hardware. NI LabVIEW software is employed to implement deterministic real-time control of the micropositioning system. A sampling rate of 5 kHz is adopted in the experimental investigations.

7.6.2 Statics Performance Testing

The statics performance of the XY stage in both the x- and y-axes has been tested by conducting experimental studies.

The two sensor outputs in each working axis are calibrated first. By applying a sinusoidal voltage signal with frequency of 0.1 Hz and amplitude of 0.5 V to the driver of VCM #1, the stage output displacement in the x-axis is measured by the laser sensor as shown in Fig. 7.7(a). The output voltages of the two strain gauges are depicted in Fig. 7.7(b). It is observed that the strain gauges $x1$ and $x2$ produce voltage ranges of 5.2087×10^{-4} and 1.0702×10^{-5} V, respectively. The two strain-gauge sensors are calibrated by comparing their output voltages to the laser displacement sensor readings. The calibrated sensors provide the output displacements as shown in Fig. 7.7(c). It is clear that strain sensor $x2$ is noisier than strain sensor $x1$ for the output displacement measurement.

Similarly, driving VCM #2 using a sinusoidal voltage signal with frequency of 0.1 Hz and amplitude of 0.5 V, the stage output displacement in the y-axis is measured by the sensors as shown in Fig. 7.8. It is observed that the strain sensor $y2$ is noisier than the strain sensor $y1$ for displacement measurement.

The resolutions of the two strain sensors in each working axis are limited by their noises. The noise mainly comes from the electric noise of the digital-to-analog converter. The motion of

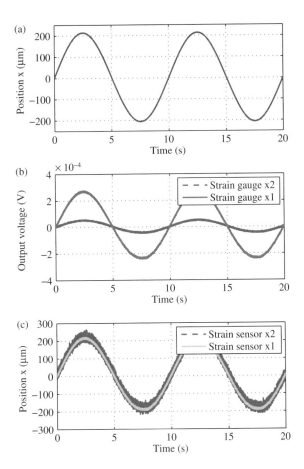

Figure 7.7 Calibration results of the x-axis strain sensors: (a) output position measured by laser sensor; (b) voltage outputs of two strain-gauge sensors $x1$ and $x2$; (c) position output of calibrated strain sensors $x1$ and $x2$

the micropositioning stage lower than the sensor noise level cannot be detected by the sensor. With a zero-voltage input, the noises of the two x-axis strain sensors are recorded as shown in Fig. 7.9(a) and (c), respectively. The histograms of the two sensor noises are depicted in Fig. 7.9(b) and (d), respectively. It is observed that the noises follow the normal distribution closely with standard deviations (σ) of 1.625 and 8.517 μm, respectively. To quantify the noise level, the σ value can be adopted as the sensor resolution [6]. Then, the resolution ratio of the coarse and fine sensors is derived as 5.24, which is 6.4% lower than the analytical prediction of 5.6. This discrepancy mainly arises from the fabrication errors of the stage parameters and the mounting errors of the strain gauges. The experimental results reveal that the resolution in the smaller stroke has been improved 5.24 times compared with that in the larger motion stroke.

Similarly, the noise-testing results of the two strain sensors in the y-axis are shown in Fig. 7.10. By adopting the standard deviations (σ) as the sensor resolutions, it is observed

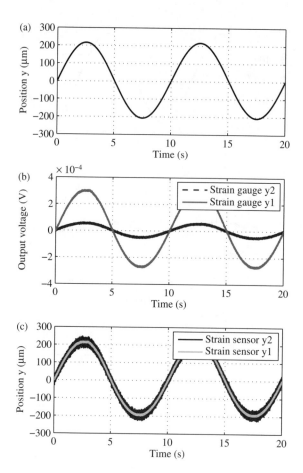

Figure 7.8 Calibration results of the y-axis strain sensors: (a) output position measured by laser sensor; (b) voltage outputs of two strain-gauge sensors y1 and y2; (c) position output of calibrated strain sensors y1 and y2

that the two sensors deliver resolutions of 1.569 and 8.421 μm, respectively. Thus, the resolution in the smaller stroke has been improved 5.37 times compared with that in the larger motion stroke for the y-axis motion.

Next, the magnitudes of the smaller and larger motion ranges are tested by applying a sinusoidal signal with an amplitude of 7.5 V, as shown in Fig. 7.11(a). The output displacements of the two x-axis strain sensors are shown in Fig. 7.11(b) and (c), respectively. It is observed that the strain sensor x1 saturates at the boundaries of the interval [−230 μm, 225 μm], which represents the smaller motion stroke of the stage. In the larger stroke of [−2452 μm, −230 μm] and [225 μm, 2554 μm], the output displacement is measured by the strain sensor x2 alone, which provides a worse resolution than the sensor x1 as tested earlier. The overall motion range [−2452 μm, 2554 μm] is constrained by the actual clearance parameter δ_2 between the mover x2 and the stroke stopper x2. Similarly, the two motion ranges in the y-axis have also been tested, and the results are shown in Table 7.3 later.

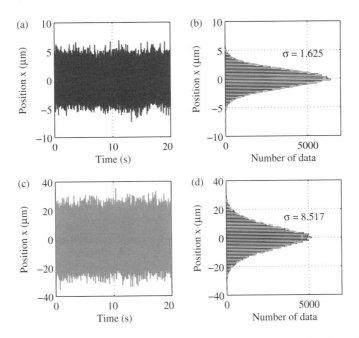

Figure 7.9   Noise-testing results of the x-axis strain sensors: (a) the measured noises of strain sensor $x1$ and (b) the noise histogram; (c) the measured noises of strain sensor $x2$ and (d) the noise histogram

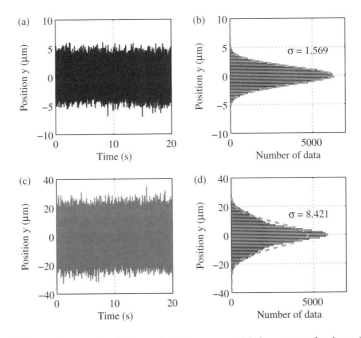

Figure 7.10   Noise-testing results of the y-axis strain sensors: (a) the measured noises of strain sensor $y1$ and (b) the noise histogram; (c) the measured noises of strain sensor $y2$ and (d) the noise histogram

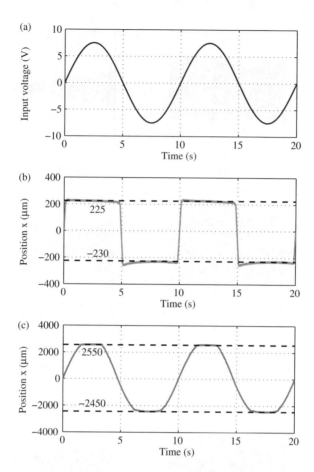

Figure 7.11 Experimental results of the x-axis motion strokes: (a) input voltage; (b) smaller motion stroke measured by strain sensor $x1$; (c) larger motion stroke measured by strain sensor $x2$

7.6.3 Dynamics Performance Testing

The dynamics performance of the XY stage is examined by the frequency response method. To examine the stage performance in the x-axis, a swept-sine signal with amplitude of 0.03 V and frequency range of 1–500 Hz is applied to drive VCM #1. The frequency responses of the stage output position in the x-axis are shown in the Bode plots in Fig. 7.12, which are generated from the outputs of the laser sensor and two strain sensors.

Owing to a small amplitude of motion, all three sensors are able to capture the first two modes at 17.5 and 29.0 Hz, respectively. Compared with the FEA simulation results of 35.2 and 94.3 Hz for the first two modes in the x-axis working direction, the experimental results are much lower. The discrepancy between the simulation and experimental results for the resonant frequencies mainly comes from the added mass of the connected mover and the moving coil of the VCM, which is not considered in FEA simulations. The resonant frequency can be enhanced by reducing the mass of the moving components.

Table 7.3   Desired and actual performances of the XY micropositioning stage

Performance	Desired x- and y-axes	Actual x-axis	Actual y-axis
Smaller stroke (mm)	[−0.2, 0.2]	[−0.225, 0.230]	[−0.210, 0.223]
Larger stroke (mm)	[−2.5, −0.2] & [0.2, 2.5]	[−2.452, −0.225] & [0.230, 2.554]	[−2.560, −0.210] & [0.223, 2.575]
Coarse/fine resolution ratio	5.6	5.24	5.32

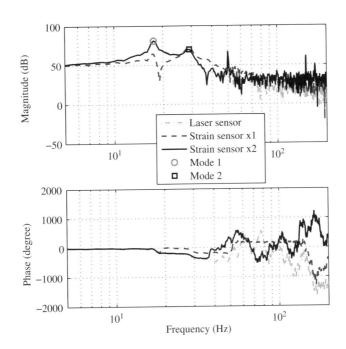

Figure 7.12   Bode plots of the x-axis frequency responses of the XY stage obtained by three sensors

While both the laser sensor and the strain sensor $x2$ predict the first mode of resonance at 17.5 Hz as the dominant resonant mode, the strain sensor $x1$ detects the second mode of resonance at 29.0 Hz as the dominant one. The reason can be seen by inspecting the two mode shapes as shown in Fig. 7.5(a) and (d), respectively. It is observed that the strain induced in the second mode is much larger than that in the first mode at the mounting place of sensor $x1$. As a result, strain sensor $x1$ recognizes the second resonant mode as the dominant mode alternatively. In contrast, the correct dominant resonant mode can be assessed by strain sensor $x2$ and the laser displacement sensor.

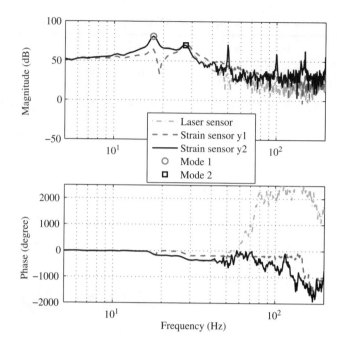

Figure 7.13   Bode plots of the *y*-axis frequency responses of the XY stage obtained by three sensors

Similarly, the frequency responses of the stage output position in the *y*-axis are shown in the Bode plots in Fig. 7.13. It is observed that all three sensors capture the first two modes at 17.5 and 27.5 Hz, respectively.

7.6.4   Circular Contouring Testing

To demonstrate the cooperative positioning performance of two axes for the XY stage, circular contouring test have been conducted. In order to achieve a precise positioning, two identical proportional-integral-derivative (PID) controllers are realized for the *x*- and *y*-axes due to the similar dynamics performance in the two working axes. The control gains of the two PID controllers are: $K_P = 4.0 \times 10^{-5}$, $K_I = 1.4 \times 10^{-5}$, and $K_D = 2.5 \times 10^{-4}$.

For example, the contouring results for a circle of 50 μm radius are shown in Figs. 7.14 and 7.15. It is observed that the XY micropositioning stage is able to provide precision circular contouring, although the two sets of strain sensors deliver different sensing resolutions.

7.6.5   Discussion

The conducted experiments validate the effectiveness of the proposed conceptual design of a multi-stroke, multi-resolution micropositioning stage with a single drive for each working axis. For a better understanding of the results, the desired and actual main performances of the XY stage are compared as shown in Table 7.3, where the results of both *x*- and *y*-axes are presented. It is observed that the smaller strokes in the two axes are larger than the design

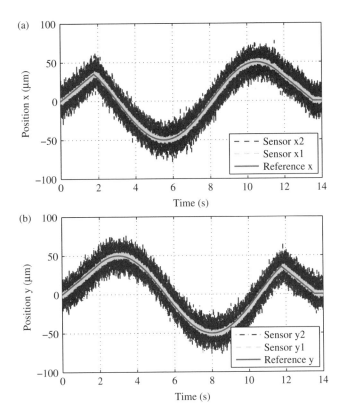

Figure 7.14   Reference and actual trajectories for the *x*-axis (a) and *y*-axis (b) of the XY stage

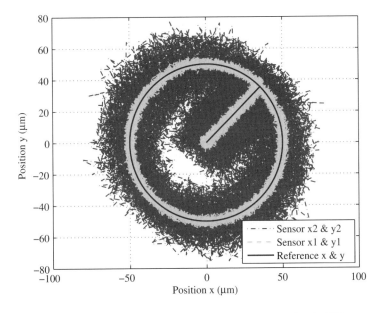

Figure 7.15   The 50 μm-radius circular contouring results for the XY stage

objective of ± 0.2 mm. The overall motion ranges in both axes are slightly larger than the design specification of 5.0 mm.

In addition, the phenomenon that both the smaller and larger bidirectional motion strokes in each axis are not exactly symmetric with respect to zero is attributed to the manufacturing tolerances and unequal clearances between the movers and each side of the stroke stoppers. The discrepancy between the desired and actual resolution ratios for the two strokes in each axis comes from the fabrication errors as well as the variation on the attached adhesive layers of the two strain gauges.

In the presented case study, the larger stroke is 10 times greater than the smaller stroke, and the positioning resolution in the smaller stroke in about 5 times higher than that in the larger stroke. The presented micropositioning system offers fine resolution only in the smaller stroke around the center of the workspace. At the limit of the smaller stroke, the impact between the mover and the stopper may cause vibration of the flexures. The impact can be alleviated by applying a low-speed reference input as depicted in the experimental results, shown in Fig. 7.11. An input signal of higher frequency will induce clear impact vibration. Even so, unlike the interference in a dual-servo stage, where the interaction is exhibited during the entire stroke of the fine stage, the impact effect only exists at the two limits of the smaller stroke in the presented stage. To eliminate the impact, a trajectory with low positioning speed at the two limits can be planned.

It is notable that the proposed design poses certain limitations in terms of bandwidth for the coarse and fine positioning. In the conventional multi-servo stage, the fine positioner exhibits a much higher bandwidth than the coarse positioner. In the reported stage, the fine motion stroke is generated by carrying the full mass of the stage, which limits the bandwidth as a consequence. Hence, a low-speed motion is expected for the smaller stroke. The proposed multi-stroke stage can be employed in potential applications which require a fine and slow-speed alignment around the home position and a coarse and fast-speed positioning in the larger stroke. For example, in microgripper-based assembly tasks, the proposed stage can be used to precisely align the microgripper to grasp an object initially, and then quickly transport the object to the destination.

Concerning the sensing issue, multi-resolution positioning is implemented using resistive position sensors in this work. Actually, there are position sensors which can provide better resolution at fine strokes. For example, the laser displacement sensors used deliver better resolution than the strain gauges. However, the sensor head and controller are bulky, which precludes their use in a compact system. The same issue exists in other fine sensors, such as the laser interferometer. In addition, the capacitance-based sensors are more compact, measuring the displacement between two parallel plates. However, a related drawback lies in the relatively small measuring range. In this work, the resistive strain gauges have been employed because they enable an embedded and compact design. Experimental results show that they are able to provide different resolutions for different strokes.

The resolutions of the strain sensors are tested using quarter-bridge circuits. The sensor resolutions can be improved by adopting half- or full-bridge circuits. Moreover, the resolution of the strain sensors can be further enhanced by removing high-frequency noises using low-pass filters. Other types of sensors, such as piezoresistive sensors, can also be exploited to improve the positioning resolutions. The presented ideas can be extended to the design of other types of multi-stroke micropositioning stages, such as rotary stages [7].

7.7   Conclusion

The conceptual design and verification of a single-drive compliant micropositioning system with multiple strokes and resolutions in each working axis has been reported in this chapter. Based on the concept of a variable-stiffness mechanism, a proof-of-concept design of a flexure-based XY stage is presented. Analytical models have been established to predict the motion strokes, coarse/fine resolution ratio, and driving force and stroke, which have been verified by FEA simulations and experimental studies. Experimental results show that the developed stage is able to produce multiple strokes using a single linear motor, and the same kind of strain-gauge sensor is capable of providing fine and coarse resolutions in the smaller and larger strokes, respectively. In the future, the design methodology will be extended to include parallel connection of the flexure bearings to provide different types of motion.

References

[1] Xu, Q. (2012) Design and development of a flexure-based dual-stage nanopositioning system with minimum interference behavior. *IEEE Trans. Automat. Sci. Eng.*, **9** (3), 554–563.

[2] Leang, K.K., Shan, Y., Song, S., and Kim, K.J. (2012) Integrated sensing for IPMC actuators using strain gauges for underwater applications. *IEEE/ASME Trans. Mechatron.*, **17** (2), 345–355.

[3] Anderson, J.K., Howell, L.L., Wittwer, J.W., and McLain, T.W. (2006) Piezoresistive sensing of bistable micro mechanism state. *J. Micromech. Microeng.*, **16** (5), 943–950.

[4] Xu, Q. (2012) New flexure parallel-kinematic micropositioning system with large workspace. *IEEE Trans. Robot.*, **28** (2), 478–491.

[5] Wheeler, A.J. and Ganji, A.R. (2009) *Introduction to Engineering Experimentation*, Prentice Hall Englewood Cliffs, NJ, 3rd edn.

[6] Dong, J., Salapaka, S.M., and Ferreira, P.M. (2008) Robust control of a parallel-kinematic nanopositioner. *J. Dyn. Sys., Meas., Control*, **130** (4), 041 007.

[7] Xu, Q. (2013) Design and implementation of a novel rotary micropositioning system driven by linear voice coil motor. *Rev. Sci. Instrum.*, **84** (5), 055 001.

Part Three

Large-Range Rotational Micropositioning Systems

8

Rotational Stage with Linear Drive

Abstract: This chapter presents the design, fabrication, and control of a compliant rotary micropositioning stage with a large rotational range. To overcome the challenge of achieving both a large rotational range and a compact size, the idea of multi-stage compound radial flexure (MCRF) is proposed. A compact rotary stage is devised to deliver a more than 10° rotational range. Analytical models are derived to facilitate the parametric design, which is validated by conducting finite-element analysis (FEA). A prototype is fabricated by employing a linear voice coil motor (VCM) and two laser displacement sensors. Proportional-integral-derivative (PID) feedback control is implemented to achieve precise rotary positioning. Experimental results demonstrate a resolution of 2 μrad over 10° rotational range and a low level of center shift of the rotary micropositioning system.

Keywords: Rotational micropositioning, Flexure mechanisms, Compliant mechanisms, Rotary stages, Multi-stage compound radial flexures, Finite-element analysis, Linear voice coil motor, Motion control.

8.1 Introduction

Limited-angle rotary micropositioning stages are required in precision engineering applications where ultrahigh-precision rotational motion within a restricted range is needed. For instance, the rotary stage can be adopted to construct rotating circular microchambers, which enable the enhancement of DNA microarray hybridizations [1]. It can work as a rotating stage for image spectrography in a multi-spectral range camera for the construction of three-dimensional architectural models [2]. In addition, it can be used as a rotary sample holder for a fluorescence microscope, which allows the generation of images along different directions [3].

In addition to the requirements on rotational range, a stage with a compact size is needed to operate in a limited space. Moreover, a compact stage allows cost reductions in terms of materials and machining processes. Unfortunately, the majority of existing rotary stages cannot achieve both a large rotational range and a compact physical size at the same time. The reason lies in the fact that these two criteria are conflicting conditions in practice. Hence, it is challenging to devise a flexure rotary system possessing both a large rotational range and a compact size simultaneously.

Design and Implementation of Large-Range Compliant Micropositioning Systems, First Edition. Qingsong Xu.
© 2016 John Wiley & Sons Singapore Pte Ltd. Published 2016 by John Wiley & Sons Singapore Pte Ltd.

To overcome the aforementioned issue, a new concept of MCRF is reported in this chapter to design a rotary stage with the merits of both a large working range and a compact dimension at the same time. In particular, the stage is driven by a linear VCM and its output motion is measured by a laser displacement sensor along with a triangular calculation. Compared with other smart material-based actuators such as piezoelectric stack actuators (PSAs) [4], the VCM offers a large stroke at the cost of a relatively small blocking force [5]. Thus, suitable actuation and sensing schemes are addressed to facilitate the stage design procedure.

For the purpose of parametric design, analytical models for performance assessment of the stage are derived, and verified by FEA simulations. Experimental results show that the developed stage produces a large rotational range over 10°. The negligible center shift allows precision rotational motion around the rotation center, which is enabled by the implemented PID feedback control. Moreover, it has a compact size which facilitates successful operation inside a limited space.

8.2 Design of MCRF

8.2.1 Limitation of Conventional Radial Flexures

To achieve rotary motion, the radial flexure as shown in Fig. 8.1(a) has been popularly used to construct a rotary stage [6], as depicted in Fig. 8.1(b). However, the rotational range of this kind of stage is limited due to the overconstraint and stress stiffening effect, which is imposed by the primary stage with fixed size.

To enlarge the rotational range, the compound radial flexures (CRFs) as shown in Fig. 8.2(a) and (b) can be employed. For example, a rotary flexure bearing has been proposed [7] by using three CRFs.

In order to achieve a large rotational range, it will be shown later that the flexures should be designed with a greater length, smaller thickness, and larger outer radius. However, these physical parameters are restricted by the compactness requirement, manufacturing process, and minimum-stiffness requirement in practice. Hence, it is practically difficult to achieve a large rotational range while maintaining a compact stage size by using the CRFs. To resolve this issue, the MCRF concept is proposed below to devise a rotary stage with an enlarged rotary angle and a compact dimension at the same time.

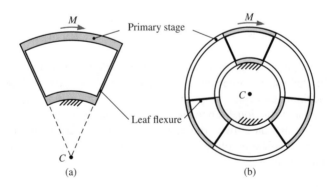

Figure 8.1   (a) A radial flexure; (b) a rotary stage composed of three radial flexures.

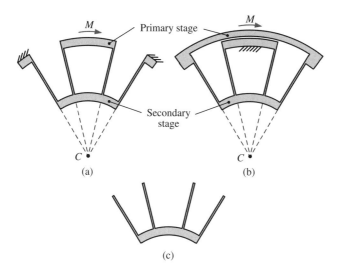

Figure 8.2 (a), (b) Compound radial flexures; (c) basic module of compound radial flexures.

8.2.2 *Proposal of MCRF*

To enlarge the rotational range, the MCRF concept is proposed, constructed by stacking N basic modules as shown in Fig. 8.2(c). To illustrate the idea, $N = 2$ is adopted to construct two MCRFs as shown in Fig. 8.3(a) and (b). By selecting the first one as an example, its performance is analyzed as follows.

With reference to Fig. 8.4, when an external moment M is applied at the primary stage of the MCRF, the primary stage rotates around the remote center point C. Due to the same length l of the eight leaf flexures, these flexures experience an identical magnitude of deformation. Moreover, a pure rotation is generated by the primary stage without parasitic motions. Instead, parasitic translations toward the primary stage along the radial direction are experienced by two secondary stages. In the following discussion, analytical models are established for the performance evaluation of the stage.

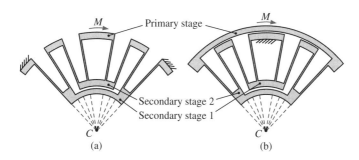

Figure 8.3 Multi-stage compound radial flexures composed of two basic modules.

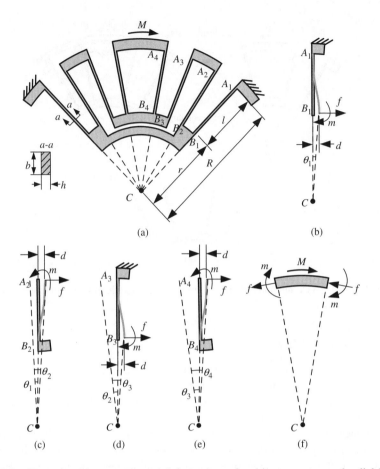

Figure 8.4   Illustration of parameters and deformations of multi-stage compound radial flexures.

8.2.3   Analytical Models

8.2.3.1   Rotary Angle

To calculate the rotary angle of the primary stage, the deformations of the four flexures A_iB_i (for $i = 1$ to 4) are shown in Fig. 8.4(b)–(e). Referring to Fig. 8.4(b), it is observed that the free end B_1 of the flexure A_1B_1 suffers from a moment m as well as a force f along the tangential direction. Hence, the free end translates a linear displacement d along the tangential direction. Due to the effect of translation d, the free end B_1 can be considered to rotate around the center point C by an angle θ_1.

In view of the boundary conditions in terms of the rotational angle and translational displacement of the free end B_1, the following relationships can be derived:

$$0 = \frac{fl^2}{2EI} - \frac{ml}{EI} \tag{8.1}$$

$$d = \frac{fl^3}{3EI} - \frac{ml^2}{2EI} \tag{8.2}$$

where E is the Young's modulus of the material and $I = bh^3/12$ is the area moment of inertia of the cross-section about the neutral axis as shown in Fig. 8.2(a).

The moment m can be solved from Eq. (8.1) as follows:

$$m = \frac{fl}{2}. \tag{8.3}$$

Owing to the small magnitude of rotation, the angle θ_1 can be approximated by the translational displacement d as follows:

$$\theta_1 = \frac{d}{r}. \tag{8.4}$$

With reference to the deformation of the flexure A_2B_2 as shown in Fig. 8.4(c), it is observed that the deflection of A_1B_1 leads to a rotation θ_1 of the entire flexure A_2B_2. In addition, the free end A_2 also experiences a tangential translation d. Hence, the total rotary angle of the end A_2 of flexure A_2B_2 around point C can be calculated as follows:

$$\theta_2 = \theta_1 + \frac{d}{R}. \tag{8.5}$$

Similarly, considering the deformations of another two flexures A_3B_3 and A_4B_4 as shown in Fig. 8.4(d) and (e), respectively, the rotary angles are derived below:

$$\theta_3 = \theta_2 + \frac{d}{r} \tag{8.6}$$

$$\theta_4 = \theta_3 + \frac{d}{R}. \tag{8.7}$$

Substituting Eqs. (8.4), (8.5), and (8.6) into Eq. (8.7) results in

$$\theta_4 = 2d \left(\frac{1}{R} + \frac{1}{r} \right). \tag{8.8}$$

Then, substituting Eq. (8.2) into Eq. (8.4) along with consideration of Eq. (8.3) allows the calculation of the rotary angle:

$$\theta_4 = \frac{fl^3(R+r)}{6EIRr} \tag{8.9}$$

which represents the one-sided rotary angle of the MCRF with two basic modules, i.e., $N = 2$. Regarding a general MCRF, the one-sided rotary angle can be deduced as follows:

$$\theta = \frac{Nfl^3(R+r)}{12EIRr} \tag{8.10}$$

where N ($N \geq 1$) is the number of basic modules. It is notable that $N = 1$ represents the special case of conventional CRF.

8.2.3.2 Torsional Stiffness

The free-body diagram of the primary stage of the MCRF is shown in Fig. 8.4(f). Considering the equilibrium of the moment around the center point C, the following equation can be derived:

$$2fR - 2m - M = 0. \tag{8.11}$$

Solving the external moment M from Eq. (8.11) and taking into account the expression (8.3) leads to

$$M = 2fR - fl. \tag{8.12}$$

Then, in view of the expressions in Eqs. (8.12) and (8.10), the torsional spring constant of a general MCRF can be calculated as follows:

$$K = \frac{M}{\theta} = \frac{12EIRr}{Nl^3} \tag{8.13}$$

where $N \geq 1$ is the number of basic modules.

8.2.3.3 Stress Analysis

If the maximum moment m_{max} is exerted by the flexures, the maximum stress σ_{max} occurs at the outermost edge of the cross-section. The maximum stress is determined by the yield strength σ_y of the material, and it can be calculated as

$$\sigma_{max} = \frac{m_{max}h}{2I} \tag{8.14}$$

which leads to the expression of the maximum moment:

$$m_{max} = \frac{2\sigma_{max}I}{h}. \tag{8.15}$$

Then, solving the force f from Eq. (8.3) and substituting it into Eq. (8.12) results in

$$M = \frac{2m(2R - l)}{l}. \tag{8.16}$$

Substituting Eq. (8.15) into Eq. (8.16) yields the maximum external moment exerted on the radial flexure:

$$M_{max} = \frac{4\sigma_{max}I(2R - l)}{lh}. \tag{8.17}$$

In view of Eqs. (8.13) and (8.17), the maximum one-sided rotary angle of the MCRF can be derived as follows:

$$\theta_{max} = \frac{M_{max}}{K} = \frac{N\sigma_{max}l^2(R + r)}{3ERrh} \tag{8.18}$$

where the relation $r = R - l$ is used.

Therefore, the maximum overall rotational range of the MCRF can be calculated as follows:

$$\Theta_{max} = 2\theta_{max} = \frac{2N\sigma_{max}l^2(R+r)}{3ERrh}. \tag{8.19}$$

Equation (8.19) shows that the maximum rotary motion of the MCRF is governed by the length l and thickness h of the leaf flexures and the radius R for a given material. To obtain a larger angle Θ_{max}, the conventional CRF ($N = 1$) should be designed with a greater length, smaller thickness, and larger outer radius. However, the physical parameters l and R are practically restricted by the compactness requirement of the stage, and h is limited by the manufacturing process and minimum-stiffness requirement. Therefore, constructed by leaf flexures with the same material and physical parameters l, h, and R, the maximum rotary angle of the proposed MCRF is enlarged N times compared with the conventional CRF. Based on the concept of MCRF, a rotary stage is designed to achieve a large rotational motion in the subsequent section.

It is notable that the leaf flexure hinge exhibits a high stress concentration at the root, where the hinge is connected to rigid connectors. In order to reduce the stress concentration factor, a fillet can be added at each root of the flexure to construct a corner-filleted hinge. In practice, the flexures are manufactured using wire-electrical discharge machining (EDM). The plate material is cut by a metal wire (e.g., molybdenum, tungsten, brass, etc.), which acts as an electrode. Usually, the wire has a diameter of 0.1 to 0.3 mm. Due to the size effect of the wire, corner-filleted flexure hinges with corner radius of 0.1 to 0.3 mm will be produced, reducing the stress concentration effect of the actual flexures. In the foregoing analysis, only the dominant bending stress of the leaf flexure is calculated. The calculation procedure of the maximum stress of a corner-filleted flexure hinge can be seen in the literature [8].

8.3 Design of a Rotary Stage with MCRF

As an embodiment of the MCRF idea, a MCRF is constructed by selecting $N = 2$. Figure 8.5(a) and (b) shows two rotary stages, which are composed of three MCRFs. For

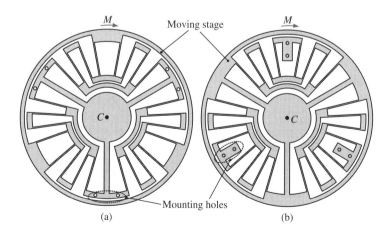

Figure 8.5 Rotary stages constructed by three MCRFs with $N = 2$.

illustration, the design of the first stage is outlined below by taking into account the actuation and sensing issues.

It is notable that $N = 3$ or more can also be used to construct the MCRFs. More or less than three MCRFs may be adopted to fabricate the rotary stage, and the design procedures are almost identical to those presented in this section.

8.3.1 Consideration of Actuation Issues

To produce a large rotary angle, a VCM is used to drive the rotary stage along the direction of F as denoted in Fig. 8.6(a). Assume that the rotational range of the stage is specified as $\pm\Phi_{max}$. The maximum one-sided angle Φ_{max} should stay within the reachable maximum angle so as to guarantee the safety of the material. That is,

$$\Phi_{max} \leq \frac{2\sigma_{max} l^2 (R + r)}{3ERrh} \tag{8.20}$$

which is generated from Eq. (8.18) with $N = 2$.

The linear VCM actuator consists of a permanent magnetic stator and a moving coil. The moving coil is a cylinder which is encompassed by a coil of copper wire, as shown in Fig. 8.6(a). The selection of a VCM with appropriate force and stroke capabilities is addressed below.

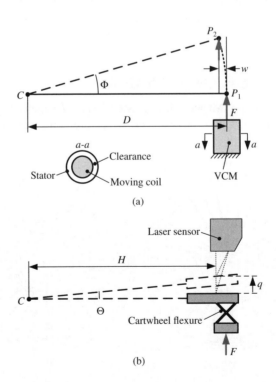

Figure 8.6 Illustrations of (a) actuating and (b) sensing principles of the rotary stage.

8.3.1.1　Force Requirement

Considering that the rotary stage is composed of three MCRFs which are connected in parallel, the torsional stiffness of the stage can be calculated via Eq. (8.13) with $N = 2$:

$$K_{\text{stage}} = 3K_\theta = \frac{18EIRr}{l^3}. \tag{8.21}$$

To generate the maximum one-sided rotary angle Φ_{max}, the required maximum force from the VCM is determined as

$$F_{\text{max}}D = K_{\text{stage}}\Phi_{\text{max}} \tag{8.22}$$

where D is the driving distance as shown in Fig. 8.6(a).

Hence, the maximum driving force of the VCM should satisfy the condition

$$F_{\text{max}}^{\text{VCM}} \geq F_{\text{max}} = \frac{K_{\text{stage}}\Phi_{\text{max}}}{D} = \frac{18EIRr\Phi_{\text{max}}}{l^3 D}. \tag{8.23}$$

8.3.1.2　Stroke Requirement

With reference to Fig. 8.6(a), the one-sided rotary angle Φ_{max} is related to the one-sided stroke S_{max} of the VCM as follows:

$$S_{\text{max}} = D \tan(\Phi_{\text{max}}). \tag{8.24}$$

Hence, to ensure that the desired rotational range is achievable, the stroke of the VCM should meet the following requirement:

$$S_{\text{max}}^{\text{VCM}} \geq 2S_{\text{max}} = 2D \tan(\Phi_{\text{max}}). \tag{8.25}$$

8.3.1.3　Driving Distance

To guarantee a proper drive of the rotary stage along the same direction by tolerating the rotary motion of the stage, the VCM is connected to the stage through a cartwheel flexure, as shown in Fig. 8.6(b), rather than using a rigid connection. To produce a rotary angle Φ, the moving coil of the VCM exhibits a lateral displacement w. Because the stator of the VCM is fixed at the base, this lateral displacement can be tolerated by the radial clearance δ between the moving coil and the stator, as shown in Fig. 8.6(a). Practically, this clearance is very small for a VCM. Hence, to ensure proper actuation, the driving distance D needs to be carefully designed.

Referring to Fig. 8.6(a) and (b), when the stage is driven to produce the maximum one-sided angle Φ_{max}, the induced lateral translation should satisfy the condition

$$w = D[1 - \cos(\Phi_{\text{max}})] \leq \delta \tag{8.26}$$

which allows the determination of the driving distance:

$$D \leq \frac{\delta}{1 - \cos(\Phi_{\text{max}})}. \tag{8.27}$$

Therefore, the conditions in Eqs. (8.20), (8.23), (8.25), and (8.27) provide guidelines for designing the stage parameters and selecting the VCM actuator to guarantee the stage rotational range of $\pm\Phi_{\text{max}}$ and the safety of the material.

8.3.2 Consideration of Sensing Issues

The rotary angle is measured by a linear optical sensor in this work. Hence, a suitable sensing scheme needs to be devised for the rotary angle measurement.

The proposed measuring principle is shown graphically in Fig. 8.6(b). The laser displacement sensor is fixed a known sensing distance H away from the rotation center C of the rotary stage. The linear displacement q of the sensing target is measured by the sensor. Then, the rotary angle can be calculated by the triangular equation

$$\Theta = \arctan\left(\frac{q}{H}\right) \tag{8.28}$$

expressed in radians.

8.4 Performance Evaluation with FEA Simulation

In this work, a rotary stage is designed to produce a rotational range of $\pm 5°$. Based on the criteria as expressed by Eqs. (8.20), (8.23), (8.25), and (8.27), the stage parameters are designed as shown in Table 8.1, where the specifications of the VCM actuator are also presented. In addition, the main specifications of the adopted Al-7075 alloy material are shown in Table 3.2.

8.4.1 Analytical Model Results

The analytical model in Eq. (8.18) predicts that the maximum one-sided rotary angle of the stage is 15.6°. Hence, the reachable rotational range of the stage is $\pm 15.6°$.

By assigning the rotational range as $\pm 5°$, the driving distance should meet the condition $D \leq 99.9$ mm to ensure lateral displacement of the moving coil lower than the radial clearance of $\delta = 0.38$ mm. By selecting a distance of $D = 58$ mm, the required maximum force and stroke from the VCM are $F_{max} = 7.73$ N and $2S_{max} = 10.15$ mm to drive the stage, determined by Eqs. (8.23) and (8.25), respectively.

Table 8.1 Main parameters of a rotary micropositioning stage

Parameter	Symbol	Value	Unit
Length of leaf flexure	l	26.0	mm
In-plane width of leaf flexure	h	0.5	mm
Out-of-plane thickness of plate material	b	10.0	mm
Outer radius of MCRF	R	50.0	mm
Deriving distance	D	58.0	mm
Sensing distance	H	60.0	mm
Stroke of the VCM motor	S_{max}^{VCM}	12.7	mm
Maximum force of the VCM motor	F_{max}^{VCM}	194.6	N
Radial clearance of the VCM motor	δ	0.38	mm

8.4.2   FEA Simulation Results

8.4.2.1   Statics Analysis

To verify the accuracy of the established models, FEA simulations are carried out using the ANSYS software package. In order to evaluate the statics performance of the rotary stage, a static structural analysis is conducted by applying a force F at the input end. Using a driving force of 1 N, the FEA result is obtained as shown in Fig. 8.7.

It is observed that the required input displacement is 0.504 mm and the produced rotary angle is 0.50°. Considering the distance D, the torsional stiffness is derived as 6.67 N·m/rad. In addition, the induced maximum stress is 13.98 MPa. Due to the isotropic property of the material, the stage performance can be deduced from these FEA results. That is, the maximum rotational range is ±17.9°. To obtain a rotational range of ±5°, the experienced maximum stress is derived as $\sigma_{max} = 139.8$ MPa. Hence, the material has a large safety factor of $\sigma_y/\sigma_{max} = 3.6$. In addition, the required maximum force and stroke from the VCM are derived as shown in Table 8.2.

Taking the FEA results as the benchmark, the errors of the analytical models are calculated as shown in Table 8.2. It is seen that the analytical models are more conservative in that they underestimate the rotary range and the required force by 12.7% and 23.0%, respectively. Yet, the established model accurately predicts the requirement of the actuator stroke, which is less than 0.25% deviation from the FEA result. The model errors mainly come from the

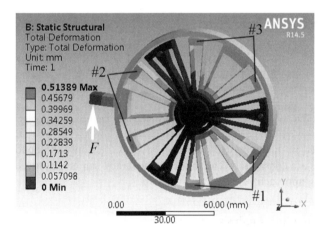

Figure 8.7   Static structural FEA simulation result for the deformation of the rotary stage.

Table 8.2   Performances of the rotary stage evaluated by analytical models and FEA simulations

Performance	Symbol	Model	FEA	Model error (%)
Maximum range (°)	$\pm\Phi_{max}$	±15.6	±17.8	−12.7
Force requirement (N)	F_{max}	7.73	10.04	−23.0
Stroke requirement (mm)	$2S_{max}$	10.15	10.12	0.25

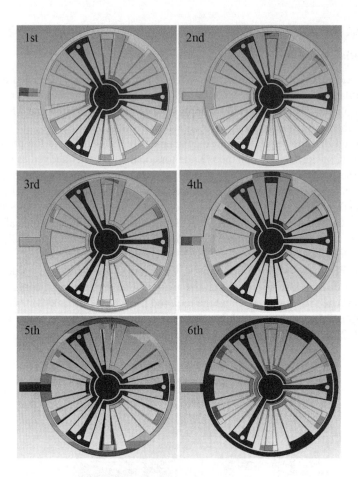

Figure 8.8   The first six mode shapes of the original design of rotary stage.

assumption adopted in the analytical modeling process, i.e., only the bending deformations of the leaf flexures are considered. The modeling accuracy can be improved by resorting to a complete modeling approach [9].

8.4.2.2   Load Capability Analysis

In practice, the mechanical load can be mounted on top of the outer frame, which acts as the output platform of the rotary stage, as shown in Fig. 8.7. By applying an external force on top of the output platform, the out-of-plane payload capability of the stage is examined. In particular, to keep the out-of-plane displacement below 50 μm, FEA simulation results predict that a payload of 1.0 kg can be supported by the stage.

8.4.2.3   Dynamics Analysis

Modal analysis is conducted to examine the dynamics performance of the rotary stage. The FEA simulation results of the first six mode shapes are shown in Fig. 8.8. The corresponding

Table 8.3 The first six resonant frequencies of the rotary stage

Mode no.	Original design (Hz)	Improved design (Hz)	Improvement (%)
1	43.91	45.94	4.6
2	134.97	161.72	19.8
3	135.13	162.16	20.0
4	168.20	170.91	1.6
5	186.38	364.39	95.5
6	211.49	408.21	93.0

resonant frequencies are tabulated in Table 8.3. It is found that the first mode arises from the dominant rotational motion. The second and third ones are the in-plane translations along two orthogonal axes, which are contributed by the multiple secondary stages. In addition, the fourth and fifth mode shapes are caused by the out-of-plane motion, and the sixth one is the in-plane rotation of secondary stages.

8.4.3   Structure Improvement

An overview of the performance evaluation reveals that the rotary stage possesses a low out-of-plane payload capability as well as a low resonant frequency for the out-of-plane motion. These performances need to be enhanced for practical applications.

Recalling the deformation shape as shown in Fig. 8.7, it is observed that the secondary stages, which are marked with the same numbers, rotate by an identical magnitude of angle. Hence, these sets of stages can be linked together, respectively, without influencing the motion property of the primary stage. To enhance the out-of-plane payload capability of the stage, the stage structure is improved by connecting these secondary stages together using three out-of-plane connecting links.

8.4.3.1   Load Capability Analysis

The out-of-plane payload capability of the improved stage is examined by FEA simulation. Results show that a payload of 3.2 kg can be tolerated by the stage output platform in order to maintain the out-of-plane displacement below 50 μm. Hence, compared with the initial design, the improved one has triple payload capability, which is contributed by an enhanced out-of-plane stiffness.

8.4.3.2   Dynamics Analysis

The resonant frequencies of the improved stage are tested by modal analysis FEA. The first six mode shapes are shown in Fig. 8.9, and the resonant frequencies are given in Table 8.3. The results reveal that the first mode shape is the in-plane rotation with an enhanced resonant frequency of 45.94 Hz. Additionally, different from the original design, the second to fourth mode shapes are attributed to the in-plane rotation of the connecting bars. The fifth and sixth

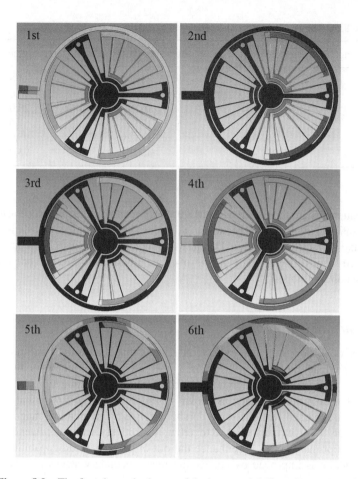

Figure 8.9   The first six mode shapes of the improved design of rotary stage.

ones are contributed by the out-of-plane motion. Hence, compared with the original design, the improved stage has a much higher out-of-plane resonant frequency; that is, 364.39 Hz versus 168.20 Hz.

Even though the connecting links add extra mass to the moving components, the improved stage exhibits a 4.6% higher resonant frequency for the dominant rotational mode. Therefore, the improved stage is more desirable for practical applications.

To validate the stage performance of the improved design, a prototype rotary stage is presented and experimental studies are carried out in the subsequent section.

8.5   Prototype Development and Experimental Studies

8.5.1   Prototype Development

Figure 8.10 shows a prototype of the rotary micropositioning stage. The rotary stage is fabricated from Al-7075 alloy by the wire-EDM process. The stage has compact dimensions, with

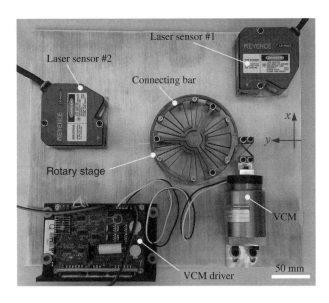

Figure 8.10   A prototype of the rotary micropositioning stage.

a diameter of 100 mm and a thickness of 10 mm. Concerning the actuator, a VCM (model: NCC05-18-060-2X, from H2W Techniques, Inc.) is chosen to provide sufficient driving force and stroke, as shown in Table 8.1. The VCM is driven by a linear power amplifier (model: LCAM7-9-ASSY, from H2W Techniques, Inc.). The VCM driver enables the selection of current or voltage driving mode. In this work, the default current mode is chosen and the electric components are adjusted to provide a gain of 2 A/V.

The rotary output motion is detected by a non-contact laser displacement sensor #1 (model: LK-H055, from Keyence Corp.) with a sub-micrometer resolution within the measurement range of 20 mm. Another laser sensor #2 is used to measure the center shift magnitudes of the stage in two perpendicular directions (in the x- and y-axes) as depicted in Fig. 8.10. In addition, an NI cRIO-9075 real-time controller (from National Instruments Corp.) equipped with NI-9263 analog output module and NI-9870 digital interface module is adopted to produce analog excitation signals and acquire digital sensor readings, respectively. LabVIEW software is employed to implement real-time control of the rotary positioning system.

8.5.2   Open-Loop Performance Testing

8.5.2.1   Statics Performance Testing

The rotational range of the stage is examined first. By applying a 0.5-Hz sinusoidal voltage signal with an amplitude of 0.75 V to the VCM driver, a current signal of 1.5-A amplitude is produced to actuate the VCM. The output position of the stage is measured by laser sensor #1. The rotary angle is derived as shown in Fig. 8.11(a), which is calculated by resorting to Eq. (8.28). It is observed that a rotational range of 10.8° is achieved by the developed stage.

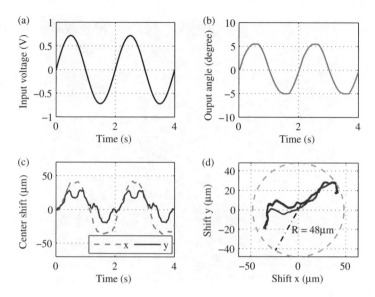

Figure 8.11 Experimental results of the rotary stage: (a) input voltage; (b) output angle; (c) center shift in x and y directions; (d) center shift in xy-plane.

As the stage rotates, the magnitudes of center shift in the x and y directions are measured by laser sensor #2, and the results are shown in Fig. 8.11(b). It is observed that the maximum shifts are 41.78 and 28.66 μm in the x and y directions, respectively. Moreover, the overall center shift in the xy-plane is plotted in Fig. 8.11(c). It is seen that the maximum radial shift is less than 48 μm, which accounts for 0.45% of the motion range of the stage. Hence, the rotary stage produces a negligible level of shift for the rotation center.

8.5.2.2   Dynamics Performance Testing

The dynamics performance of the stage is examined by the frequency response method. Specifically, a swept-sine signal with amplitude of 0.05 V and frequency range of 1–200 Hz is applied to the VCM driver. The frequency responses of the rotary motion are shown in Fig. 8.12, generated by the fast Fourier transform (FFT) algorithm.

It is observed that the stage exhibits a resonant frequency at 15.0 Hz, which is lower than the frequency of 45.9 Hz as predicted by FEA. The discrepancy between the experimental and FEA simulation results mainly arises from the added mass of the cartwheel flexure and the moving coil of the VCM, which is not considered in FEA simulations. The resonant frequency can be enhanced by choosing a VCM with smaller mass of the moving coil.

8.5.3   Controller Design and Closed-Loop Performance Testing

To achieve a precise positioning, a PID controller is implemented and the closed-loop performance of the micropositioning stage is tested. A block diagram of the control system is shown in Fig. 8.13.

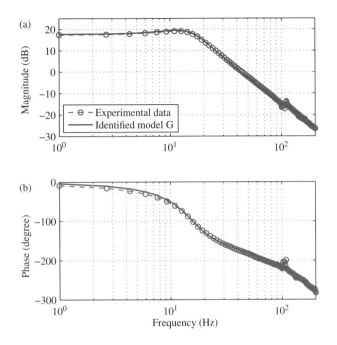

Figure 8.12   Open-loop frequency responses of the rotary stage: (a) magnitude plot; (b) phase plot.

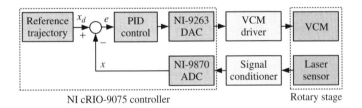

Figure 8.13   Block diagram of the rotary micropositioning control system.

8.5.3.1   PID Controller Design

Owing to its popularity, the traditional PID controller is implemented as follows:

$$u(t) = K_c \left[e(t) + \frac{1}{T_i} \int_0^t e(\tau) d\tau + T_d \frac{de(t)}{dt} \right] \tag{8.29}$$

where t is the time variable, the displacement error $e(t) = x_d(t) - x(t)$ with x_d and x representing the desired and actual system output, respectively. In addition, K_c, T_i, and T_d denote the controller gain, integral time, and derivative time, respectively.

The PID control parameters are tuned through simulation study with Matlab software. The control simulation is conducted based on a fourth-order model (8.30), which is identified from the experimental results of the frequency responses as shown in Fig. 8.12:

$$G(s) = \frac{0.052s^4 - 60.52s^3 + 9.256 \times 10^4 s^2 - 3.555 \times 10^7 s + 4.382 \times 10^{10}}{s^4 + 189.5s^3 + 6.485 \times 10^5 s^2 + 5.603 \times 10^7 s + 5.735 \times 10^9}. \tag{8.30}$$

Specifically, based on the plant model (8.30), the PID control gains are tuned by the Ziegler–Nichols (Z–N) method through extensive simulation studies. During the parameter tuning process, a sampling time of 0.004 s is selected. The tuning procedures are shown below.

(a) First, T_i and T_d are set as infinity and zero, respectively.
(b) Then, K_c is increased gradually until K_c attains K_u, which excites an oscillating output with a period of T_u as well as unchanged amplitude.
(c) Based on the values of K_u and T_u, PID parameters are selected as $K_c = 0.6K_u$, $T_i = 0.5T_u$, and $T_d = 0.125T_u$. That is, $K_c = 0.0930$, $T_i = 0.0103$ s, and $T_d = 0.0026$ s.

8.5.3.2 Positioning Performance

First, the set-point positioning performance of the rotary micropositioning system is examined using the PID controller. For a 1° step input, the results are shown in Fig. 8.14(a). It is observed that the PID control produces a 5% settling time of 0.078 s without overshoot. Hence, it enables the stage to have a rapid transient response. In addition, a steady-state root-mean-square error (RMSE) of 1.226 μrad is yielded by the rotary stage. Moreover, a consecutive-step positioning with step size of 2 μrad is carried out, and the results are shown in Fig. 8.14(b). The fact that the step size can be clearly identified indicates that the rotary positioning resolution of the system is better than 2 μrad.

Next, the motion tracking of a 0.5-Hz sinusoidal trajectory with 5° amplitude is carried out. The tracking results are depicted in Fig. 8.15, which covers almost the entire rotational range of the micropositioning system. It is observed that the trajectory is followed accurately with a RMSE of 0.152°; that is, 1.5% of the motion range. In addition, the relationship between the output angle and the reference input is shown in Fig. 8.16. It is seen that the closed-loop hysteresis is mitigated to 3.8% of the motion range. Compared with the open-loop result of 9.6%, the PID control has significantly mitigated the hysteresis effect by 60%.

Then, the frequency responses of the closed-loop micropositioning system with PID controller are evaluated. By applying a sinusoidal trajectory with amplitude of 0.5° and frequency ranging from 0.1 to 30 Hz, the Bode plots of the PID control system are generated as shown in Fig. 8.17. It is seen that a large phase lag over 90° exists within the ordinary −3 dB bandwidth, which leads to large tracking errors. Alternatively, the closed-loop control bandwidth is defined as the frequency at which the phase is lagged 30°. Using the PID control, a 30°-lag bandwidth of 5.25 Hz is obtained, which is equivalent to 35% of the resonant frequency of the positioning system. Due to a relatively low bandwidth of the system, the tracking error for a high frequency input is large. As future work, a damping control will be implemented to attenuate the resonant peak, and the control bandwidth of the micropositioning system can be broadened accordingly.

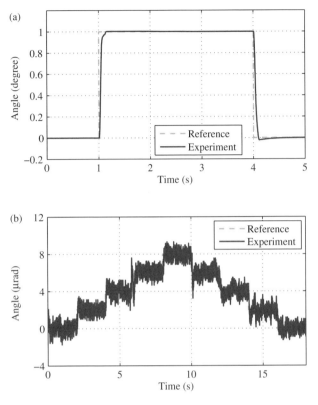

Figure 8.14   Rotary positioning results of (a) 1° step rotation response and (b) 2-μrad consecutive step response.

8.5.4   *Further Discussion*

It is notable that the readings of sensor #2 in the statics testing are a combination of the center shift displacement and the surface roughness of the output stage. To verify the effectiveness of the employed sensing method and to eliminate the effect of surface roughness, the center shift value can be measured by using a computer vision-based approach instead. In addition, the magnitude of the closed-loop positioning error as shown in Fig. 8.15 indicates that the positioning accuracy of the rotary stage can be further enhanced. Even so, the PID control is capable of mitigating the hysteresis effect and leading to a 2-μrad positioning accuracy for the rotary stage.

For a clear comparison, the performances of existing typical compliant rotary micropositioning stages in terms of the planar dimension, positioning range, and resolution are listed in Table 8.4. It is observed that the rotary stage reported in this work delivers the largest rotational range and possesses the most compact size. It is notable that a motion range of 35° can be attained as predicted by FEA simulation.

Concerning the laser displacement sensor, it is claimed by the supplier that the measurement will not be influenced by the inclination of the measurement target provided that the reflected light is detected by the sensor. The positioning resolution relies on the performance of the employed displacement sensor. In future, displacement sensors with higher resolution

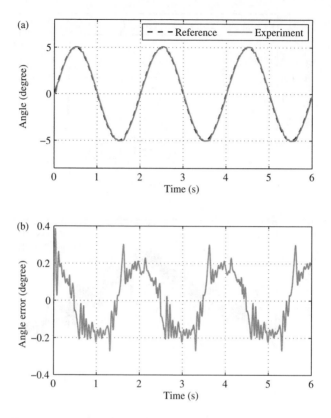

Figure 8.15   (a) Sinusoidal tracking results and (b) tracking errors of the rotary stage with PID control.

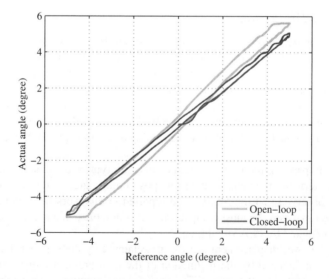

Figure 8.16   Open-loop and closed-loop hysteresis loops of the rotary stage.

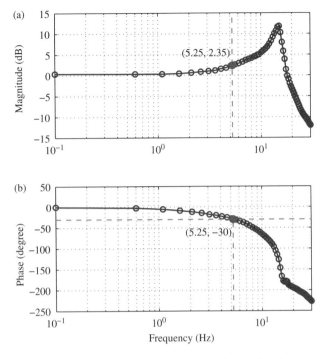

Figure 8.17 Closed-loop frequency response of the rotary micropositioning system: (a) magnitude plot; (b) phase plot.

Table 8.4 Performance comparison of typical rotary micropositioning stages

Reference	Outer diameter (mm)	Rotational range (°)	Resolution (μrad)
[10]	≈120	0.006	0.13
[11]	≈120	0.06	0.50
[6]	104	0.3	—
[12]	158	0.4	0.01
This work	100	10.8	2.00

and lower noise will be adopted to improve the positioning resolution of the rotary micropositioning system.

The rotary stage is applicable to precision engineering where a precise rotary positioning within a restricted angle is required. In further research, a more compact rotary stage will be produced by resorting to an optimum architectural design for related applications.

8.6   Conclusion

A compact rotary micropositioning stage with a large rotational range is designed and developed in this chapter. Based on the concept of MCRF, a rotary stage is devised as an example.

Analytical models are established to predict the torsional stiffness, maximum rotary angle, and required motor stroke and force, verified by FEA. Furthermore, the stage performances are examined by several experimental studies on a fabricated prototype. Both simulation and experimental results indicate that the stage is capable of rotary positioning with a resolution of 2 μrad over a 10° range, while possessing a compact size. Moreover, the large rotary angle is achieved with a negligible level of shift of the rotation center. Advanced control schemes will be realized to accomplish a more precise rotary positioning for pertinent applications.

References

[1] Vanderhoeven, J., Pappaert, K., Dutta, B., Hummelen, P.V., and Desmet, G. (2005) DNA microarray enhancement using a continuously and discontinuously rotating microchamber. *Anal. Chem.*, **77** (14), 4474–4480.

[2] Brusco, N., Capeleto, S., Fedel, M., Paviotti, A., Poletto, L., Cortelazzo, G., *et al.* (2006) A system for 3D modeling frescoed historical buildings with multispectral texture information. *Mach. Vision Appl.*, **17** (6), 373–393.

[3] Greger, K., Swoger, J., and Stelzer, E.H.K. (2007) Basic building units and properties of a fluorescence single plane illumination microscope. *Rev. Sci. Instrum.*, **78** (2), 023–705.

[4] Yao, Q., Dong, J., and Ferreira, P.M. (2007) Design, analysis, fabrication and testing of a parallel-kinematic micropositioning XY stage. *Int. J. Mach. Tools Manuf.*, **47** (6), 946–961.

[5] Tan, K.K., Huang, S., Liang, W., Mamun, A.A., Koh, E.K., and Zhou, H. (2012) Development of a spherical air bearing positioning system. *IEEE Trans. Ind. Electron.*, **59** (9), 3501–3509.

[6] Kim, K., Ahn, D., and Gweon, D. (2012) Optimal design of a 1-rotational DOF flexure joint for a 3-DOF H-type stage. *Mechatronics*, **22** (1), 24–32.

[7] Valois, M., Rotary flexure bearing. US Patent No. 0034027A1 (Feb. 9, 2012).

[8] Lobontiu, N., Paine, J.S.N., Garcia, E., and Goldfarb, M. (2001) Corner-filleted flexure hinges. *J. Mech. Des.*, **123**, 346–352.

[9] Xu, Q. and Li, Y. (2011) Analytical modeling, optimization and testing of a compound bridge-type compliant displacement amplifier. *Mech. Mach. Theory*, **46** (2), 183–200.

[10] Wang, Y.C. and Chang, S.H. (2006) Design and performance of a piezoelectric actuated precise rotary positioner. *Rev. Sci. Instrum.*, **77** (10), 105–101.

[11] Hwang, D., Byun, J., Jeong, J., and Lee, M.G. (2011) Robust design and performance verification of an in-plane XYθ micropositioning stage. *IEEE Trans. Nanotechnol.*, **10** (6), 1412–1423.

[12] Shu, D., Lee, W.K., Liu, W., Ice, G.E., Shvyd'ko, Y., and Kim, K.J. (2011) Development and applications of a two-dimensional tip-tilting stage system with nanoradian-level positioning resolution. *Nucl. Instrum. Methods Phys. Res. Sect. A-Accel. Spectrom. Dect. Assoc. Equip.*, **649** (1), 114–117.

9

Rotational Stage with Rotary Drive

Abstract: This chapter presents the design, modeling, and experimental testing of a flexure-based compliant rotational micropositioning stage driven by a rotary voice coil motor (VCM). The rotational stage is devised based on multi-stage compound radial flexures (MCRFs). It has both a compact size and a large rotary stroke. To cater for the requirements of rotational angle and driving torque, the dominant parameters of the stage are determined based on the established analytical models. These models are verified by conducting finite-element analysis (FEA) simulations. A prototype rotary stage is developed by employing a strain-gauge angle sensor. The sensitivities of the angle sensor are derived analytically and calibrated experimentally. Results show that the developed compact stage is capable of rotary positioning with a resolution of 20 μrad over the range of 10° under an approximately infinite fatigue life.

Keywords: Rotational micropositioning, Flexure mechanisms, Compliant mechanisms, Rotary stages, Multi-stage compound radial flexures, Finite-element analysis, Rotary voice coil motor, Strain-gauge sensor, Fatigue failure analysis.

9.1 Introduction

The rotary stage as reported in Chapter 8 is able to deliver a large rotational angle. However, it exhibits several shortcomings. First, it is driven by a linear VCM along the tangential direction of the output platform and its rotational angle is limited by the clearance between the moving coil and the stator of the VCM. Second, the rotational angle of the stage is measured by a laser displacement sensor, which makes the overall device bulky and unsuitable for applications in a limited space.

To overcome the aforementioned issues, a new large-range rotary stage driven by a rotary VCM is presented in this chapter. To facilitate the parametric design, analytical models are established to predict the stage performance, which is verified by conducting FEA simulations. The rotational angle of the stage is measured by a designed strain-gauge sensor. The proposed scheme enables a compact design with embedded angle sensing. Moreover, a prototype rotary stage is developed for experimental testing. Experimental results indicate that the reported rotary flexure stage is capable of delivering a larger rotational angle compared with previous work.

9.2   New Design of MCRF

In this section, a new MCRF module is proposed to facilitate the design of compliant rotary stages.

9.2.1   MCRF Design

As shown in Fig. 9.1(a), a basic radial flexure module is proposed to devise CRFs and MCRFs as illustrated in Fig. 9.1(b), (c) and (d), (e), respectively. Without loss of generality, the MCRFs are constructed using two basic modules (i.e., $N = 2$), although more basic modules can also be adopted. It is notable that the CRFs consist of one basic module (i.e., $N = 1$), where N is the number of basic modules. Hence, the CRF can be considered as a special case of the MCRF with $N = 1$.

Taking the MCRF as shown in Fig. 9.1(d) as an example, its performance is analyzed in the subsequent section. When an external moment M is applied at the primary stage of the MCRF,

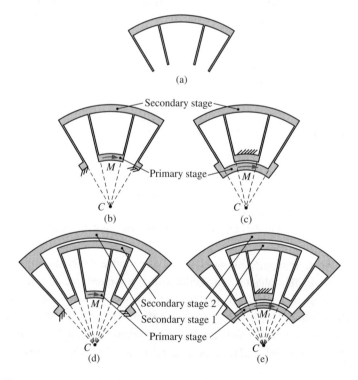

Figure 9.1   The proposed radial flexures: (a) basic module; (b), (c) compound radial flexures; (d), (e) multi-stage compound radial flexures.

the primary stage rotates around the remote center point C. Because of the same length l of the eight leaf flexures, these flexures experience an identical magnitude of deformation. Moreover, a pure rotation is generated by the primary stage without parasitic motions. Instead, parasitic translations toward the primary stage along the radial direction are borne by the two secondary stages. To facilitate a performance evaluation of the stage, analytical models are established in the following discussion.

9.2.2 Analytical Model Not Considering Deformation

First, the symbolic formulation as presented in the literature [1] is employed for the analytical compliance/stiffness evaluation of the rotary stage. The formulation is developed based on screw theory. It is useful to describe the relationship between the motion twists and force wrenches of flexure elements. By this approach, the compliance/stiffness matrix of an entire flexure mechanism can be generated by operating the compliance/stiffness matrices of every flexure hinge through appropriate coordinate transformations. The calculation procedure is presented as follows.

Concerning a leaf flexure as shown in Fig. 9.2, its motion twist T and the force wrench W can be related by a 6×6 compliance matrix C or stiffness matrix K as follows:

$$T = CW \tag{9.1}$$

$$W = KT \tag{9.2}$$

where

$$T = [\theta_x, \theta_y, \theta_z, \delta_x, \delta_y, \delta_z]^T \tag{9.3}$$

$$W = [F_x, F_y, F_z, M_x, M_y, M_z]^T \tag{9.4}$$

$$K = C^{-1}. \tag{9.5}$$

When the compliance matrix C is expressed in the coordinate system located at the flexure end, it can be written as follows [1]:

$$
C = \begin{bmatrix}
0 & 0 & 0 & \frac{l}{GJ_x} & 0 & 0 \\
0 & 0 & -\frac{l^2}{2EI_y} & 0 & \frac{l}{EI_y} & 0 \\
0 & \frac{l^2}{2EI_z} & 0 & 0 & 0 & \frac{l}{EI_z} \\
\frac{l}{EA} & 0 & 0 & 0 & 0 & 0 \\
0 & \frac{l^3}{3EIz} & 0 & 0 & 0 & \frac{l^2}{2EI_z} \\
0 & 0 & \frac{l^3}{3EI_y} & 0 & -\frac{l^2}{2EI_y} & 0
\end{bmatrix}
\tag{9.6}
$$

where E and G are the Young's modulus and shear modulus of the material, respectively. In addition, $A = bh$, $I_y = b^3 h/12$, $I_z = bh^3/12$, and $J_x = bh(b^2 + h^2)/12$.

Referring to Fig. 9.3, the connection relations of the eight leaf flexures can be described by

$$(1 \parallel 2) \ddagger (3 \parallel 4) \ddagger (5 \parallel 6) \ddagger (7 \parallel 8) \tag{9.7}$$

where $\parallel$ and $\ddagger$ represent the parallel and serial connections, respectively.

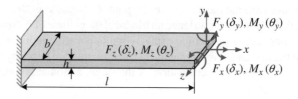

Figure 9.2   Parameters and coordinate system of a leaf flexure hinge.

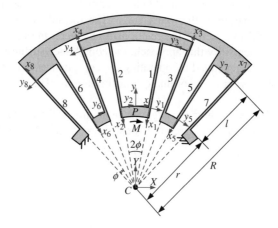

Figure 9.3   Coordinate systems of compliance matrices of eight flexure hinges for a MCRF.

The stiffness matrices of parallel elements can be added together and the compliance matrices of serial elements can be added together, as long as these matrices are expressed in a common coordinate system. Hence, after transforming the compliance matrices of the eight hinges from their local coordinates to the common coordinate system at point P, the compliance matrix of a MCRF can be written in the coordinate system at point P as follows:

$$C_p = \left(T_1 K T_1^{-1} + T_2 K T_2^{-1}\right)^{-1} + \left(T_3 K T_3^{-1} + T_4 K T_4^{-1}\right)^{-1}$$

$$+ \left(T_5 K T_5^{-1} + T_6 K T_6^{-1}\right)^{-1} + \left(T_7 K T_7^{-1} + T_8 K T_8^{-1}\right)^{-1} \tag{9.8}$$

where the transformation matrix T_i (for $i = 1$ to 8) is expressed as

$$T_i = \begin{bmatrix} R_i & 0 \\ D_i R_i & R_i \end{bmatrix} \tag{9.9}$$

with R_i being a 3×3 rotation matrix and D_i a skew-symmetric matrix defined by the vector d_i.

For example, the coordinate transformation of the flexure hinge 1 is conducted using

$$
R_1 = \begin{bmatrix} \cos\left(-\left(\dfrac{\pi}{2}+\phi\right)\right) & -\sin\left(-\left(\dfrac{\pi}{2}+\phi\right)\right) & 0 \\ \sin\left(-\left(\dfrac{\pi}{2}+\phi\right)\right) & \cos\left(-\left(\dfrac{\pi}{2}+\phi\right)\right) & 0 \\ 0 & 0 & 1 \end{bmatrix}
\tag{9.10}
$$

$$
d_1 = \begin{bmatrix} r\,\sin\phi \\ -(r - r\,\cos\phi) \\ 0 \end{bmatrix}
\tag{9.11}
$$

$$
D_1 = \begin{bmatrix} 0 & 0 & -r + r\,\cos\phi \\ 0 & 0 & -r\,\sin\phi \\ r - r\,\cos\phi & r\,\sin\phi & 0 \end{bmatrix}
\tag{9.12}
$$

where ϕ is the inclined angle between two adjacent hinges.

Then, the compliance matrix C_p is transformed as C_c, which is expressed in the coordinate system at the rotation center C. The stiffness matrix K_c of the MCRF is obtained by inverting C_c. The relation between the external torque M around the rotation center C and the rotation angle θ_z of the MCRF is described by the torsional stiffness:

$$
K_I = \frac{M}{\theta_z} = K_c(6,3)
\tag{9.13}
$$

which is the sixth row, third column element of the stiffness matrix K_c.

9.2.3   Analytical Model Considering Deformation

The foregoing method can be used to predict the torsional stiffness of the MCRF through a matrix approach without considering the deformed shape. To better understand the force and deformation status of each individual flexure hinge, an alternative method is presented as follows.

9.2.3.1   Rotational Angle Calculation

To calculate the rotational angle of the primary stage as produced by an external torque M, the deformations of the four flexures A_iB_i (for $i = 1$ to 4) as shown in Fig. 9.4(a) are depicted in Fig. 9.4(b)–(e).

Referring to Fig. 9.4(b), the free end A_1 of the flexure A_1B_1 bears a moment m and a force f along the tangential direction. Hence, the free end A_1 translates a linear displacement d along the tangential direction. Given the translation d, the free end A_1 rotates around the center point C by an angle θ_1.

Taking into account the boundary conditions in terms of the translational displacement and rotational angle of the free end A_1, the following relationships hold:

$$
\frac{fl^3}{3EI} - \frac{ml^2}{2EI} = d
\tag{9.14}
$$

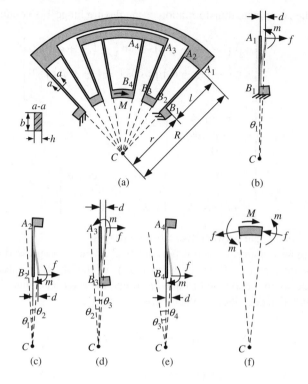

Figure 9.4   Parameters and free-body diagrams of a multi-stage compound radial flexure.

$$\frac{fl^2}{2EI} - \frac{ml}{EI} = 0 \tag{9.15}$$

where E is the Young's modulus of the material and $I = bh^3/12$ is the moment of inertia. The force f can be determined from Eq. (9.15) as follows:

$$f = \frac{2m}{l}. \tag{9.16}$$

Due to the small magnitude of rotation, the angle θ_1 can be expressed approximately by the translational displacement d:

$$\theta_1 = \frac{d}{R}. \tag{9.17}$$

Regarding the deformation of the flexure A_2B_2 as shown in Fig. 9.4(c), a rotation θ_1 of the entire flexure A_2B_2 is induced by the deflection of A_1B_1. In addition, the free end B_2 also experiences a tangential translation d. Hence, the total rotational angle of end point B_2 of the flexure A_2B_2 around the point C can be calculated as follows:

$$\theta_2 = \theta_1 + \frac{d}{r}. \tag{9.18}$$

Similarly, considering the deformations of the other two flexures A_3B_3 and A_4B_4, as shown in Fig. 9.4(d) and (e), respectively, their rotational angles are derived below:

$$\theta_3 = \theta_2 + \frac{d}{R} \tag{9.19}$$

$$\theta_4 = \theta_3 + \frac{d}{r}. \tag{9.20}$$

Substituting Eqs. (9.17), (9.18), and (9.19) into Eq. (9.20) results in

$$\theta_4 = 2d\left(\frac{1}{R} + \frac{1}{r}\right). \tag{9.21}$$

Then, substituting Eq. (9.14) into Eq. (9.21) and recalling Eq. (9.16) allows the calculation of the rotational angle:

$$\theta_4 = \frac{ml^2(R+r)}{3EIRr} \tag{9.22}$$

which represents the one-sided rotational angle of the MCRF with two basic modules (i.e., $N = 2$).

For a general MCRF, the one-sided rotational angle can be deduced as follows:

$$\theta = \frac{Nml^2(R+r)}{6EIRr} \tag{9.23}$$

where N ($N \geq 1$) is the number of basic modules. It is notable that $N = 1$ represents the case of a conventional CRF.

9.2.3.2   Torsional Stiffness Calculation

The free-body diagram of the primary stage of the MCRF is shown in Fig. 9.4(f). Considering the equilibrium of the moment around the center point C, the following equation is obtained:

$$2fr + 2m - M = 0. \tag{9.24}$$

Solving the external moment M from Eq. (9.24) and taking into account Eq. (9.16) gives

$$M = \frac{4mr}{l} + 2m. \tag{9.25}$$

Then, in view of Eqs. (9.25) and (9.23), the torsional spring constant of a general MCRF can be calculated as follows:

$$K_{\mathrm{II}} = \frac{M}{\theta} = \frac{12EIRr}{Nl^3} \tag{9.26}$$

where $N \geq 1$.

9.2.3.3   Stress Calculation

Driven by a torque M, the leaf flexures of MCRF suffer dominantly from bending deformations. If the maximum moment m_{max} is exerted on the flexures, then the maximum stress σ_{max} occurs at the outermost edge of the cross-section, which can be calculated by [2]

$$\sigma_{max} = \frac{m_{max} h}{2I} \tag{9.27}$$

to give the maximum moment

$$m_{max} = \frac{2\sigma_{max} I}{h}. \tag{9.28}$$

Substituting Eq. (9.28) into Eq. (9.25) yields the maximum external moment experienced by the MCRF:

$$M_{max} = \frac{4\sigma_{max} I(2r + l)}{lh}. \tag{9.29}$$

In view of Eqs. (9.26) and (9.29) as well as the relation of $R = r + l$, the maximum one-sided rotational angle of the MCRF can be determined as

$$\theta_{max} = \frac{M_{max}}{K_{II}} = \frac{N\sigma_{max} l^2 (R + r)}{3ERrh}. \tag{9.30}$$

Hence, the maximum overall rotational range of the MCRF is calculated as follows:

$$\Theta_{max} = 2\theta_{max} = \frac{2N\sigma_{max} l^2 (R + r)}{3ERrh} \tag{9.31}$$

where $r = R - l$.

Equation (9.31) reveals that the maximum rotational range of the MCRF is governed by the length l and thickness h of the leaf flexures as well as the radius R of the stage for a given material. To obtain a larger angle Θ_{max}, the conventional CRF ($N = 1$) should be designed with a greater length, smaller thickness, and larger outer radius. However, in practice, the physical parameters l and R are restricted by the compactness requirement of the stage. In addition, h is limited by the manufacturing tolerance and the minimum-stiffness requirement, which influences the natural frequency of the stage. Therefore, constructed by leaf flexures with identical physical parameters in terms of l, h, and R, the maximum rotational angle of the proposed MCRF is enlarged N times compared with the conventional CRF. Based on the concept of MCRF, a rotary stage is designed to achieve a large rotational motion in the following section.

9.3 Design of the Rotary Stage

By employing three MCRFs with $N = 2$, a rotary stage is devised as shown in Fig. 9.5, which exhibits two different fixing schemes of the stage. Specifically, the outer frame in the first scheme is fixed and the inner stage acts as the output platform, as shown in Fig. 9.5(a). Meanwhile, the inner stage in the second one is fixed and the outer frame is considered as the output stage, as illustrated in Fig. 9.5(b). Here, three MCRFs are used because two MCRFs may result in an out-of-plane bending deformation of the stage when an external load is exerted, while four or more MCRFs lead to smaller clearance between the flexures, hence producing a smaller rotational angle of the stage.

For illustration, the design procedures of a stage with the first fixing scheme are presented below by taking into account the actuation and sensing issues. It is notable that $N = 3$ or

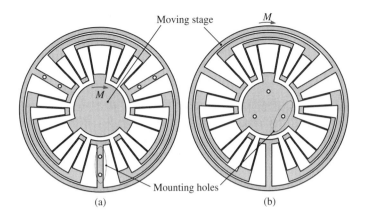

Figure 9.5 A rotary stage constructed from three multi-stage compound radial flexures: (a) the inner stage is used as output stage; (b) the outer frame is used as output stage.

more can also be used to construct the MCRFs. The design process is similar to the one presented here.

Similar to the treatment in Chapter 8, the stage structure can be improved by connecting three pairs of secondary stages using extra links to enhance the natural frequency and out-of-plane payload capability. The three links are illustrated in Fig. 9.6. One pair of secondary stages can be connected together because the two components experience identical rotational displacement. Hence, the connection does not impose constraints on the rotational output motion of the rotary stage. In addition, mechanical stoppers can be added to

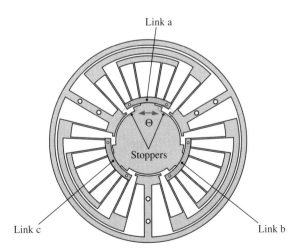

Figure 9.6 An improved design of the rotary stage where the three pairs of secondary stages are connected by links a, b, and c. Mechanical stoppers are used to protect the rotary stage from exceeding the maximum angle.

prevent the rotary stage from exceeding the desired maximum rotational angle, as shown in Fig. 9.6.

9.3.1   Actuator Selection

To produce a large rotational angle, a rotary VCM is used to drive the rotary stage to rotate around the center point C. As a rotary actuator, the VCM consists of a permanent magnetic rotor and a stator. The stator is a hollow cylinder encompassed by a coil of copper wire. To ensure proper operation, a VCM with appropriate torque and stroke capabilities is selected below.

Specifying the rotational range of the stage as $\pm\Phi_{max}$, the maximum one-sided rotation Φ_{max} should stay within the maximum one-sided reachable range as stipulated below to guarantee the safety of the material:

$$\Phi_{max} \leq \frac{2\sigma_{max} l^2 (R+r)}{3ERrh} \tag{9.32}$$

which is determined from Eq. (9.30) with $N = 2$.

Because the rotary stage is composed of three MCRFs which are connected in parallel, the stage's torsional stiffness can be calculated by using the analytical models in Eq. (9.13) or Eq. (9.26) with $N = 2$ as follows:

$$K_{stage}^{I} = 3K_{I} = 3K_c(6, 3) \tag{9.33}$$

$$K_{stage}^{II} = 3K_{II} = \frac{18EIRr}{l^3}. \tag{9.34}$$

To achieve the maximum one-sided rotational angle Φ_{max}, the required torque from the VCM motor is determined by

$$M_{max} = K_{stage}\Phi_{max}. \tag{9.35}$$

The maximum driving torque of the VCM should satisfy the following condition:

$$M_{max}^{motor} \geq M_{max}. \tag{9.36}$$

Considering that the rotary stage is driven by the VCM motor directly, the total rotational stroke of the VCM should meet the following requirement to ensure achievement of the desired rotational range:

$$\Phi_{max}^{motor} \geq 2\Phi_{max} \tag{9.37}$$

Therefore, the aforementioned conditions in Eqs. (9.32), (9.36), and (9.37) provide guidelines for the design of stage parameters and the selection of VCM to guarantee the stage's rotational range of $\pm\Phi_{max}$ and the safety of the material.

9.3.2   Sensor Design

To measure the rotational angle of the stage, a strain-gauge sensor is designed due to its compact size. To enhance the sensitivity, the strain gauge is attached at the maximum-strain

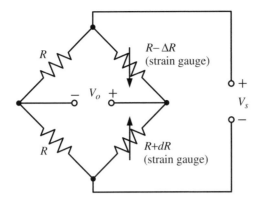

Figure 9.7   Half-bridge circuit for the signal conditioning of two strain gauges.

position of the leaf flexure [4]. The position of the maximum strain can be determined by conducting FEA simulation.

In this work, the sensor output is acquired through a half-bridge circuit as shown in Fig. 9.7. The output voltage of the circuit can be approximated by

$$V_o = \frac{V_s}{2R} \times \Delta R \qquad (9.38)$$

where ΔR and R are the change value and nominal value of the gauge resistance, respectively. V_s is the source voltage of the circuit.

For a strain gauge, the gauge factor is expressed as

$$S = \frac{\Delta R}{R\,\epsilon} \qquad (9.39)$$

in which the strain ϵ is related to the experienced stress σ by

$$\sigma = E\,\epsilon \qquad (9.40)$$

where E is the Young's modulus of the material.

Substituting Eqs. (9.39) and (9.40) into Eq. (9.38) yields

$$V_o = \frac{V_s S\sigma}{2E} \qquad (9.41)$$

which indicates that the output voltage is proportional to the experienced stress σ of leaf flexures. Hence, the strain gauge can be glued at the position of maximum stress to enhance the sensing sensitivity.

To determine the sensitivity of the strain-gauge sensor, one leaf flexure is selected for analysis as conducted below.

Referring to Fig. 9.4, the free end of each flexure undergoes a translation d in the tangential direction relative to the fixed end. The linear stiffness of the leaf flexure can be derived as follows [5]:

$$K_t = \frac{Ebh^3}{l^3}. \qquad (9.42)$$

Recalling Eq. (9.16), the following relationship holds:

$$f = \frac{2m}{l} = K_t d \tag{9.43}$$

where the torque m can be derived from Eq. (9.27):

$$m = \frac{2\sigma I}{h} \tag{9.44}$$

with σ denoting the experienced stress.

In addition, given the expressions (9.17) and (9.21), the translation d can be related to the rotational angle θ of the stage through

$$d = \frac{Rr\theta}{2(R + r)}. \tag{9.45}$$

Substituting Eqs. (9.44) and (9.45) into Eq. (9.43), a fundamental algebra operation gives

$$\sigma = \frac{K_t Rrhl\theta}{8I(R + r)}. \tag{9.46}$$

Then, the output voltage of the circuit can be generated by substituting Eq. (9.46) into Eq. (9.41):

$$V_o = \frac{V_s S K_t Rrhl\theta}{16EI(R + r)}. \tag{9.47}$$

Furthermore, substituting Eq. (9.42) and the moment of inertia $I = bh^3/12$ into Eq. (9.47) yields

$$V_o = S_{\text{angle}}\theta \tag{9.48}$$

where the sensitivity S_{angle} of the strain gauge working as an angle sensor is

$$S_{\text{angle}} = \frac{3V_s SRrh}{4l^2(R + r)}. \tag{9.49}$$

Equation (9.48) reveals that the relation between the stage's rotational angle θ and the strain-gauge circuit output V_o is linear. In addition, the expression (9.49) indicates that the angle sensor sensitivity S_{angle} is governed by the supply source voltage, the gauge factor, and the stage parameters. In practice, the sensitivity value S_{angle} of the strain sensor can be determined by means of experimental calibration.

9.4   Performance Evaluation with FEA Simulation

In this section, a rotary stage is designed to produce a rotational range of $\pm 5°$. Based on the design criteria as expressed by Eqs. (9.32), (9.36), and (9.37), the stage parameters are designed as shown in Table 9.1.

Table 9.1   Main parameters of a rotary stage

Parameter	Symbol	Value	Unit
Length of leaf flexure	l	26.0	mm
In-plane width of leaf flexure	h	0.6	mm
Out-of-plane thickness of plate material	b	10.0	mm
Outer radius of MCRF	R	39.0	mm
Inclined angle of two adjacent flexures	ϕ	12.0	°

9.4.1   Analytical Model Results

By assigning the rotational range as $\pm 5°$, the analytical model I based on symbolic formulation predicts that the required maximum torque from the VCM is $M_{max}^{motor} \geq 0.84$ N·m, which is obtained by Eqs. (9.33), (9.35), and (9.36). In addition, analytical model II results on the required maximum torque and stroke for the VCM are $M_{max}^{motor} \geq 0.58$ N·m and $\Phi_{max}^{motor} \geq 10°$, which are predicted by Eqs. (9.34), (9.35), (9.36), and (9.37).

Moreover, the analytical model II in Eq. (9.30) predicts that the maximum one-sided rotational angle of the stage is $31.0°$, which is constrained by the yield strength of the material. Hence, the reachable rotational range of the stage is $\pm 31.0°$. Compared with the reachable rotational range of $\pm 31.0°$, the assigned value of $\pm 5°$ is obtained with a high safety factor of 6.2, which is the ratio between the yield strength and the experienced maximum stress of the material.

9.4.2   FEA Simulation Results

9.4.2.1   Static FEA Results

To verify the accuracy of the established models, FEA simulations are carried out with the ANSYS software package. In order to generate better results, the 20-node element SOLID186 is selected to create the mesh model with ANSYS.

Specifically, to assess the statics performance of the stage, a static structural analysis is conducted by applying a moment of 0.01 N·m on the center stage (i.e., the output platform of the rotary stage). For instance, Fig. 9.8 shows the FEA results obtained with 74 160 elements. The produced rotational angle is $0.1017°$. Hence, the torsional stiffness is derived as 5.63 N·m/rad. The induced maximum stress is 2.22 MPa, which occurs around the inner ends of the leaf flexures as shown in Fig. 9.8(b). Due to an isotropic property of the material, the performance of the stage can be deduced from the FEA results. That is, the maximum rotational range of the stage is $\pm 23.1°$. To obtain a rotational angle of $\pm 5°$, the required maximum torque from the VCM is 0.50 N·m and the material exhibits a safety factor of 4.6.

To investigate the effect of the mesh density, FEA simulations have been conducted under different mesh sizes. For each mesh size, the analytical model errors of the required torque and the maximum rotational angle are calculated by taking FEA results as the benchmark, and the results are shown in Figure 9.9. It is observed from Fig. 9.9(a) that the more the elements, the larger the model errors of the required torque prediction by both models. Figure 9.9(b) reveals that the error of the developed analytical model II for the maximum rotational angle prediction

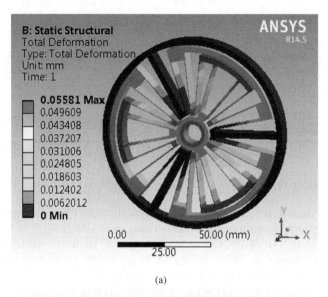

(a)

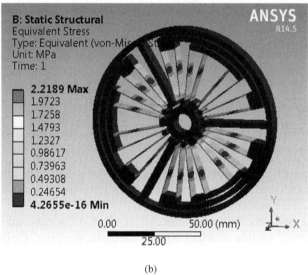

(b)

Figure 9.8   Static FEA simulation results generated by fixing the outer frame and applying a torque at the center stage: (a) result of total deformation; (b) distribution of equivalent stress.

does not change monotonically as the number of elements increases. For example, with 74 160 elements, the model error of the maximum rotational angle prediction is 34% with respect to FEA results. The torque calculation errors of the symbolic model I and the reported analytical model II are 71% and 18%, respectively.

Comparatively, the error of the symbolic model I is about threefold larger than that of the analytical model II. One possible reason is that the torsional stiffness predicted by symbolic model I only considers the external torque (M_z), while both external torque (M_z) and internal

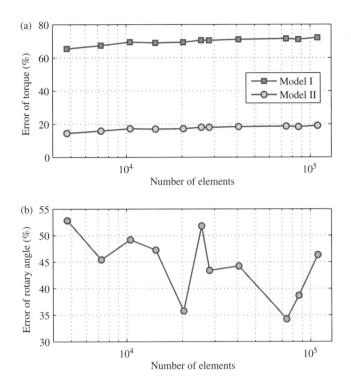

Figure 9.9 Model error versus mesh density of FEA simulation results: (a) model error of the torque requirement; (b) model error of the rotational angle. The model errors are calculated by (Model−FEA)/FEA×100%.

force (F_y) are taken into account by the developed analytical model II. Another reason may lie in the fact that model I is developed using the initial status of the rotary stage without considering the deformed shape, whereas model II is established by examining the deformed shape of each flexure hinge. As a result, analytical model II captures the angle–torque relationship of the stage more accurately than symbolic model I.

In addition, the error of analytical model II for the maximum rotational angle exceeds 30% with respect to FEA results. The discrepancy is mainly caused by the assumption used in the analytical model, where only the bending deformations are considered to calculate the maximum stress. Hence, the analytical model overestimates the rotational angle of the rotary stage. A more accurate model is desirable to capture the rotational angle of the stage.

The out-of-plane payload capability of the rotary stage is examined by applying an external force on top of the stage output platform. FEA results reveal that a payload of 15.7 kg can be carried by the stage with an out-of-plane displacement less than 500 μm. The induced maximum stress is σ_{max} = 212.0 MPa, which implies a safety factor of 2.4 for the material.

9.4.2.2 Dynamic FEA Results

The dynamics performance of the stage is evaluated by conducting the modal analysis using ANSYS software. To obtain a better assessment, the moving components including the rotor of

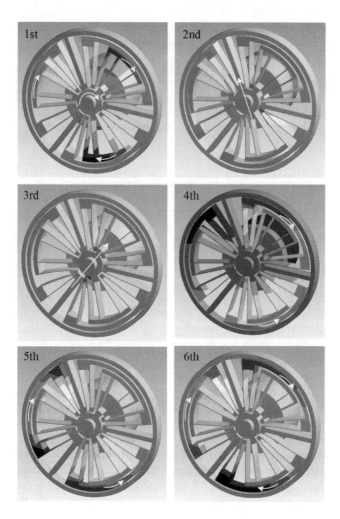

Figure 9.10   Simulation results of the first six mode shapes of the rotary stage.

the VCM are assumed to be steel material, and a total mass of 220 g is added to the output stage in the simulation study. The different parts are assigned different materials and then combined as a single part. Afterwards, it is meshed and the fixing constraints are applied in ANSYS. The first six mode shapes are shown in Fig. 9.10, and the corresponding resonant frequencies are tabulated in Table 9.2. It is observed that the first resonant mode arises from the dominant rotational motion of the output platform. The second and third ones are caused by bending of the motor with respect to the xy-plane in different directions. The fourth to sixth ones are attributed to in-plane translations or rotations of the secondary stages.

For the xy-plane comparison purposes, modal analysis is also conduced for an improved design as shown in Fig. 9.6, and the resonant frequencies are presented in the last column of Table 9.2. Concerning the rotary stage reported here, the first natural frequencies of the present and improved designs are 30.25 and 31.85 Hz, respectively. That is, the improved

Table 9.2 FEA simulation results of resonant frequencies of the rotary stage

Mode no.	Present design (Hz)	Improved design (Hz)
1	30.25	31.85
2	34.18	90.48
3	34.21	91.10
4	110.38	122.34
5	110.43	122.64
6	113.14	126.06

design exhibits a first natural frequency which is 1.6 Hz higher than the present design. A high natural frequency is desirable because the higher the first resonant frequency, the faster the device can operate in the virtual static mode. However, the achieved improvement by the improved design is not significant. Hence, such a structure improvement is not very demanding from the viewpoint of the first natural frequency. In this sense, the proposed rotary stage design simplifies the stage architecture and the fabrication process compared with previous work [3]. It is found that the second and third resonant frequencies are close to the first resonant frequency of the present design. In practice, these two resonant modes, which are caused by the motor bending with respect to the *xy*-plane, are not likely to occur because the rotor of the VCM motor is cylindrically constrained.

9.4.2.3   Fatigue-Lifecycle FEA Results

Fatigue analysis is necessary in the design phase of the rotary stage. To produce a desired rotational range of $\pm5°$, the fatigue life of the stage structure is predicted through FEA simulation.

Specifically, by applying a fully reversed torque of 0.50 N·m to the output platform of the rotary stage, stress life analysis is carried out. FEA simulation results show that the maximum number of lifecycles is about 4.9×10^6 before fatigue failure of the structure. Hence, the stage structure exhibits an approximately infinite life when producing a $\pm5°$ rotational range.

9.5   Prototype Fabrication and Experimental Testing

9.5.1   Prototype Development

Figure 9.11 shows a CAD model of the designed rotary stage, and Fig. 9.12 depicts a prototype rotary micropositioning stage. The monolithic rotary stage is fabricated using a piece of Al 7075 alloy by means of the wire-electrical discharge machining (EDM) process. The stage exhibits a compact dimension, with diameter of 100 mm. A rotary VCM is chosen for the actuation by considering the torque and stroke requirements. In particular, the VCM (model: MR-040-25-025, from H2W Techniques, Inc.) is selected to deliver sufficiently large torque and stroke. The rotational output motion of the stage is measured by two metallic strain gauges (model: SGD-3/350-LY13, from Omega Engineering Ltd.). The strain gauge has a nominal

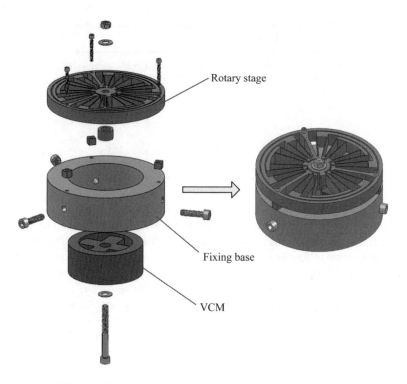

Figure 9.11　CAD model of a rotary micropositioning stage.

resistance of 350 Ω and a gauge factor of $S = 2.0$. Figure 9.8(b) reveals that the maximum stress occurs around the inner ends of the leaf flexures. Hence, the strain gauges are glued at these positions to obtain a relatively high sensitivity.

In order to measure the half-bridge circuit output, an NI-9949 half-bridge completion accessory is used to complete the 350 Ω sensor. The NI-9949 contains two high-precision resistors of 350 Ω. The bridge output is acquired by using the NI-9237 input module, which provides a 24-bit data acquisition resolution.

For calibration of the strain sensor, the rotational motion of the rotary stage is also monitored by a digital microscope with a magnification ratio of 200. The microscope captures the image of a copper wire which is attached on the rotary output platform. The rotational angle of the stage is calculated by an image processing technique. In addition, an NI cRIO-9022 real-time controller combined with NI-9118 chassis is adopted as the control hardware. LabVIEW software is employed to realize deterministic real-time control of the system using a sampling rate of 5 kHz.

9.5.2　Statics Performance Testing

First, the rotational range of the micropositioning stage is tested. By applying a step signal as shown in Fig. 9.13(a) to the VCM driver, the output voltage of the strain sensor is acquired as shown in Fig. 9.13(b). The output angles of the stage at three positions I, II, and III are

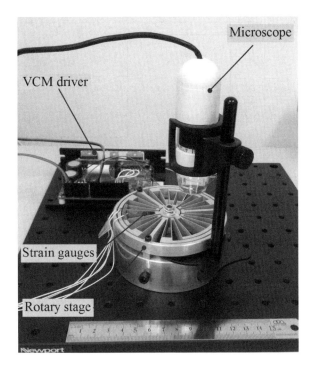

Figure 9.12   A prototype rotary micropositioning stage.

captured by the microscope. Then, the rotational angles are calculated from the overlay image of the three frames as shown in Fig. 9.14. These angles correspond to the initial position I and two limit positions II and III, as shown in Fig. 9.13(b). By setting the angle in position I as the zero initial value, the output angles in positions II and III are calculated as 5.951° and −5.002°, respectively. Hence, an overall rotational range of 10.953° is achieved by the present design.

By comparing the output voltages of the strain sensor at the limit positions (II and III) with the calculated angles, the sensor is calibrated using a gain of $1/S_{\text{angle}} = 1680.8°/V$, which is the reciprocal of the sensor sensitivity. In addition, Eq. (9.49) predicts a sensor gain value of $1/S_{\text{angle}} = 1765.6°/V$. Compared with the experimental result, the analytical prediction overestimates the sensor gain by 5%. The discrepancy may arise from the manufacturing tolerance and the difference between the actual and nominal values of the strain-gauge factor.

It is observed from Fig. 9.13(b) that some damping effect is present in the response. In view of the overshoot magnitude of the step response, the damping ratio is calculated as 0.56. For some applications, such as accelerometers, the ideal normalized damping is 0.7 because it maximizes the flatness of the amplitude response. Some other devices are designed to be critically damped with a damping ratio of 1.0. Concerning the developed rotary stage, if a desired damping ratio is needed, a closed-loop control technique can be implemented to adjust the damping of the controlled device.

The noise of the calibrated strain sensor is acquired with a zero-voltage input to the rotary stage. A second-order Butterworth filter with cutoff frequency 30 Hz is used to remove the

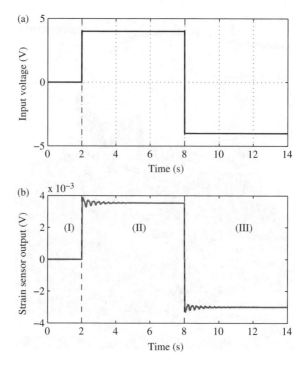

Figure 9.13   (a) Input voltage used to drive VCM; (b) strain sensor output voltage.

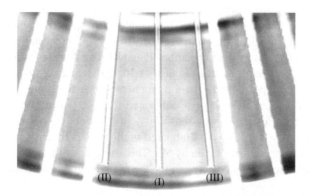

Figure 9.14   The overlay image of three frames for three rotational positions of a copper wire that is fixed on the output platform.

high-frequency components. The time history and histogram of the noise signal are shown in Fig. 9.15(a) and (b), respectively. The standard deviation is calculated as $\sigma = 0.000572°$ $= 9.98$ μrad. By adopting 2σ as the resolution, the positioning resolution of the rotary stage is determined as 19.96 μrad.

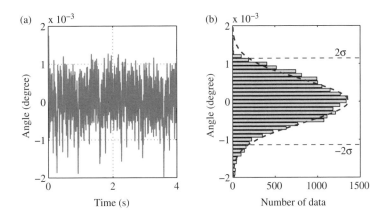

Figure 9.15   (a) Time history and (b) histogram of the noise signal of the strain-gauge angle sensor.

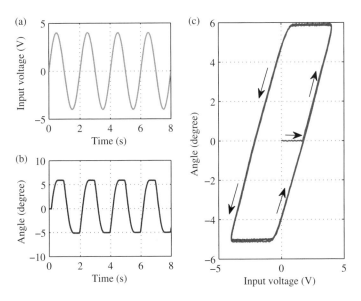

Figure 9.16   (a) Driving voltage signal; (b) output angle measured by strain sensor; (c) output angle versus input voltage of the rotary stage.

   Driving the rotary stage using a 0.5-Hz sinusoidal voltage signal as shown in Fig. 9.16(a), the output angle is measured by the strain sensor. The results are shown in Fig. 9.16(b). The relationship between the output angle and the input voltage is described in Fig. 9.16(c), which exhibits a hysteresis loop. The hysteresis mainly comes from the VCM motor, which works based on the Lorentz force law. In order to achieve a precise positioning, a control technique is needed to suppress the hysteresis nonlinearity.

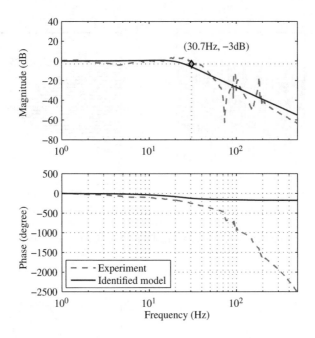

Figure 9.17 Bode plots of the rotary stage: magnitude response (upper) and phase response (lower).

9.5.3   Dynamics Performance Testing

The dynamics performance of the rotary stage is tested using the frequency response method. Specifically, a swept-sine signal with amplitude of 2 V and frequency ranging from 1 to 500 Hz is produced to drive the VCM. The rotational angle response is captured by the strain sensor. The frequency response of the rotational angle is shown in Fig. 9.17, which is generated by the fast Fourier transform (FFT) algorithm.

It is observed that the stage exhibits a weak resonant peak around 21 Hz. A second-order model with damping ratio of 0.56 is identified, and its frequency response is plotted in Fig. 9.17. It is seen that the identified model is able to capture the low-frequency dynamics up to 50 Hz. To characterize the system behavior at higher frequencies, a model of higher order is required.

9.5.4   Discussion

The foregoing experimental results reveal that the developed rotary stage is able to deliver a rotational range of 10.953°, which is larger than the results reported in existing works [3, 6, 7]. The conducted FEA simulation predicts a reachable rotational range of ±23.1° for the rotary stage. Hence, the rotational range can be further increased by using a motor with larger actual driving torque. In addition, the phenomenon that the bipolar rotational range is asymmetric with respect to zero may be caused by the manufacturing tolerance of the stage parameters and the assembly errors of the VCM and the rotary stage.

Compared with the FEA simulation results of the first resonant frequency at 30.25 Hz, the prediction results of 21 Hz from the identified second-order model and experimental result are 30% lower. This discrepancy may be induced by the fabrication errors of the stage parameters. As a consequence, the actual flexures are not as rigid as the strength of the material models employed in the FEA. The resonant frequency of the device can be enhanced by choosing a VCM with a lighter rotating rotor. In addition, Fig. 9.17 shows that the –3 dB bandwidth of the developed stage is derived as 30.7 Hz. This indicates that the maximum speed of the rotary stage is up to 30.7 Hz before there is a 3-dB attenuation of the magnitude response.

It is notable that the out-of-plane payload capability of the rotary stage can be enhanced by the improved design of the rotary stage, as shown in Fig. 9.6. FEA simulation shows that the improved stage can support an out-of-plane load of 45 kg with a safety factor of 2.4 for the material. Compared with the loading weight of 15.7 kg for the original stage without connecting bars, the load capability has been enhanced almost twofold. In addition, by linking these secondary stages, the second resonant frequency has been enhanced to be twice as high as the first one, as shown in Table 9.2, which indicates a more robust rotational motion of the rotary stage.

The positioning resolution is dependent on the performance of the sensor. Here, the strain-gauge sensor is used because it allows a compact design and enables a faster sampling rate than the visual servoing approach. The major adverse issue of using the strain-gauge sensor lies in the considerable noise. Advanced signal processing techniques are needed to filter out the noise without influencing the measured signal. Alternatively, rotary optical encoders with higher resolution and lower noise may be adopted to further improve the positioning resolution of the rotary positioning device.

The designed rotary stage is fabricated with Al alloy in this work. The rotary stage can also be miniaturized and fabricated using Si for MEMS applications. Compared with classical rotary springs (e.g., Archimedes spring or serpentine springs used in MEMS [8]), the advantage of the reported rotary flexure is that it is capable of a large rotational range with a small center shift value. The reported rotary stage is applicable to precision engineering scenarios where a precise rotary positioning over a limited angle is required. For example, the compliant rotary stage with 10° range can be applied in optical scanning applications [9] and laser beam pointing in space applications [10]. In further research, a more compact rotary stage will be produced by resorting to an optimum architectural design.

The presented design idea can also be extended to the design of large-range rotary stages with multiple axes. For instance, Fig. 9.18 illustrates a large-range compliant universal hinge, which is constructed from two MCRFs with $N = 2$. By replacing the compliant universal hinges of the 3-PUU parallel mechanism as shown in Fig. 1.7 with the proposed ones, a large-range translational motion in the three-dimensional space is easily achieved.

It is notable that only the open-loop performance of the rotary stage is tested in this chapter. In the future, a closed-loop control scheme will be implemented to achieve a precise positioning for the rotary stage.

9.6   Conclusion

This chapter reports a compact flexure-based rotary micropositioning stage with a large rotational range along with embedded angle sensing. Based on the MCRF module, a new rotary

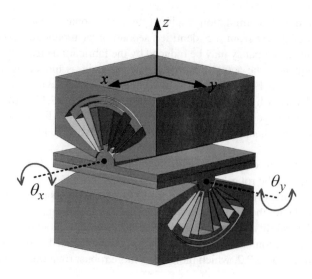

Figure 9.18   Large-range compliant universal hinge constructed from two MCRFs with $N = 2$.

stage is devised. Analytical models are established to predict the maximum rotational angle, torsional stiffness, and the required motor stroke and torque, which are validated by conducting FEA simulations. Furthermore, the stage performances are tested by conducting several experimental studies on a prototype stage. The simulation and experimental results indicate that the stage is capable of rotary positioning with a resolution better than 20 μrad over a range of 10.953°. As predicted by FEA simulations, the rotational range is generated with a large safety factor of 4.6 with a fatigue-lifecycle of 4.9×10^6 for the material, which indicates a large reachable rotational range over 45°. Benefiting from the mechanical design of the stage structure and the embedded sensor design, the rotary stage possesses a compact physical dimension. In the future, the proposed idea will be extended to the design of large-range rotary micropositioning stages with multiple axes.

References

[1] Su, H.J., Shi, H., and Yu, J. (2012) A symbolic formulation for analytical compliance analysis and synthesis of flexure mechanisms. *J. Mech. Des.*, **134** (5), 051 009.

[2] Smith, S.T. (2000) *Flexures: Elements of Elastic Mechanisms*, CRC Press, Boca Raton, FL, p. 203.

[3] Xu, Q. (2013) Design and implementation of a novel rotary micropositioning system driven by linear voice coil motor. *Rev. Sci. Instrum.*, **84** (5), 055 001.

[4] DiBiasio, C.M. and Culpepper, M.L. (2012) A building block synthesis approach for precision flexure systems with integrated, strain-based position sensing. *Precis. Eng.*, **36** (4), 673–679.

[5] Xu, Q. (2012) New flexure parallel-kinematic micropositioning system with large workspace. *IEEE Trans. Robot.*, **28** (2), 478–491.

[6] Wang, Y.C. and Chang, S.H. (2006) Design and performance of a piezoelectric actuated precise rotary positioner. *Rev. Sci. Instrum.*, **77** (10), 105 101.

[7] Kim, K., Ahn, D., and Gweon, D. (2012) Optimal design of a 1-rotational DOF flexure joint for a 3-DOF H-type stage. *Mechatronics*, **22** (1), 24–32.

[8] Tang, W.C., Nguyen, T.C.H., and Howe, R.T. (1989) Laterally driven polysikmn resonant microstructures. *Sens. Actuator A-Phys.*, **20**, 25–32.

[9] Chong, J., He, S., and Ben Mrad, R. (2012) Development of a vector display system based on a surface-micromachined micromirror. *IEEE Trans. Ind. Electron.*, **59** (12), 4863–4870.

[10] Pijnenburg, J., Rijnveld, N., and Hogenhuis, H. (2012) Extremely stable piezo mechanisms for the new gravitational wave observatory, in *Proc. SPIE 8450, Modern Technologies in Space- and Ground-Based Telescopes and Instrumentation II* (eds R. Navarro, C.R. Cunningham, and E. Prieto), Amsterdam; doi:10.1117/12.926100.

Part Four

Applications to Compliant Gripper Design

10

Large-Range Rotary Gripper

Abstract: This chapter presents the design, analysis, and testing of a robotic compliant gripper with integrated position and force sensors dedicated to automated microhandling tasks. The gripper delivers a large gripping range with a bidirectional drive. It is capable of detecting the gripping force and environmental interaction forces in horizontal and vertical axes. A variable-stiffness compliant mechanism is designed to provide the force sensing with dual sensitivities and dual measuring ranges. Analytical models are derived to predict the gripping range, force sensing sensitivities and ranges, which are verified by conducting finite-element analysis (FEA) simulations. A prototype gripper is developed for experimental studies. Results show that the single strain-gauge force sensor is able to detect the gripping and interaction forces in an alternate manner. The dual-sensitivity, dual-range force sensor provides a solution to fine and coarse force sensing in smaller and larger ranges, respectively.

Keywords: Robotic grippers, Compliant mechanisms, Rotary flexure bearing, Large-range grippers, Gripping force sensing, Environmental interaction force sensing, Dual-sensitivity force sensing, Finite-element analysis, Strain-gauge force sensors.

10.1 Introduction

A robotic gripper is a typical tool to realize an automated pick-and-place operation. It has been applied extensively in micromanipulation and microassembly tasks [1, 2]. In the literature, various sensorized grippers with different structures, actuators, and sensors have been proposed for microhandling applications. In particular, compliant mechanisms deliver attractive merits in terms of no backlash, no friction, no wear, lubrication free, and easy to manufacture. Hence, they have been widely employed in microhandling applications [3–5].

10.1.1 Structure Design and Driving Method

Depending on the applications of concern, compliant grippers can be fabricated in macro-, meso- and microscales. In the literature, both linear and rotary guiding approaches have been

Design and Implementation of Large-Range Compliant Micropositioning Systems, First Edition. Qingsong Xu.
© 2016 John Wiley & Sons Singapore Pte Ltd. Published 2016 by John Wiley & Sons Singapore Pte Ltd.

proposed in gripper structure design. However, the majority of the existing compliant grippers can only deliver a limited gripping range, typically less than 1 mm. To adapt to the gripping of objects with various sizes, it is desirable to design a gripper with a large gripping range in order to facilitate wider application. From the viewpoint of compliant mechanism design, it is challenging to devise a gripper with such a large gripping range while maintaining the overall dimensions as compact as possible [4].

Generally, the gripper arms are only able to provide a unidirectional gripping motion. For a normally open gripper, the gripper arms are usually actuated to close the gripper tips in a positive direction, while the open operation in the negative direction is realized by the restoring force provided by flexure mechanisms. In practice, the capability to drive in both positive and negative directions enables a twofold gripping range of the gripper compared with the unidirectionally driven version. However, seldom have grippers been reported to provide bidirectional drive in close and open operation of the gripper tips. One reason is that the majority of actuators can only provide one-way drive (e.g., electrostatic actuators, piezoelectric stack actuators, etc.). To implement the drive in both positive and negative directions, two actuators can be adopted [6]. Yet, this will complicate the gripper design process and increase the hardware costs. To overcome this shortcoming, an electromagnetic actuator is employed in this chapter to deliver a bidirectional drive of the gripper arm. As for the cost, the gripper has a relatively large dimension due to the large size of the actuator.

In this chapter, a compliant gripper is designed to deliver a large gripping range over 4 mm for microassembly tasks. Specifically, the gripper structure is designed using multi-stage compound parallelogram flexures (MCPFs) and multi-stage compound radial flexures (MCRFs) to generate a large-range linear and rotary guide, respectively. The role of the MCRF-based rotary guiding mechanism is to transfer the y-axis actuation displacement into the x-axis motion of the gripper arm, which enables a compact design of the gripper structure, as shown in Fig. 10.1.

10.1.2   Sensing Requirements

When the gripper arms are closed to grasp an object, the gripping force sensing is important to guarantee that an appropriate force is exerted on the object. The position and force sensing is a crucial issue to realize an accurate and reliable grasp operation. The reason lies in the fact that a small force is not sufficient to grasp the object firmly while a large force may incur damage to the object. To ensure reliable operation, it is also essential to detect the interaction force exerted by the environment [7]. Such an interaction force sensing is important to determine whether the gripper contacts the environment and to ensure the safety of the gripper device by avoiding excess interaction. The majority of existing grippers can provide the gripping force sensing in the gripping direction alone, whereas seldom have grippers been developed with the capability of interaction force sensing.

In the literature, the computer vision has been employed to detect the contact force successfully [3, 8]. However, the limitation on the depth of view of the microscope restricts its wide application. Alternatively, a thermally actuated gripper with capacitive type of gripping and interaction force sensors has been proposed [9], where two force sensors are used to detect the gripping force and interaction force along two perpendicular directions, respectively. Recently, a miniature gripper with piezoresistive sensor for the measurement of contact position and magnitude of a perpendicular external interaction force has been developed [10].

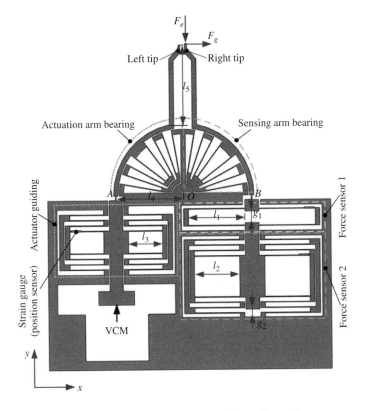

Figure 10.1   Schematic of the designed compliant gripper.

However, previous works require separate force sensors for the gripping and interaction force sensing, which complicates the gripper structure design process and increases the hardware costs.

The reason for using two force sensors mainly arises from the fact that a force sensor can only generally measure the force in a single direction. It is challenging to devise a force sensor for the detection of the applied forces in two perpendicular directions. Moreover, many practical applications require a two-stage force sensing; that is, a high-sensitivity sensing in a smaller range and a low-sensitivity sensing in a larger range to execute different tasks. Usually, a force sensor can only provide a single sensitivity within a specific measuring range. Designing a force sensor with dual sensitivities and ranges poses another challenge.

In the subsequent sections of this chapter, the design of a large-range gripper with integrated position and force sensing is presented. Specifically, a new structure design is proposed to enable a large gripping range over 4 mm for the left tip of the gripper. A dual-sensitivity, dual-range force sensor is designed to measure both gripping and interaction forces in perpendicular directions for the right tip. The idea of unequal stiffness as presented in Chapter 6 is further extended and adapted to force sensing to generate a strain gauge-based force sensor, which provides a dual-sensitivity force sensing in dual ranges. The force sensor flexure is connected to the gripper arm through a compliant rotary bearing. In this way, the gripping and interaction forces applied to the gripper arm are detected by the same sensor sequentially.

10.2 Mechanism Design and Analysis

In this work, leaf flexures are employed to yield a large elastic deformation. The mechanical structure of the gripper is designed as shown in Fig. 10.1. The compliant gripper consists of a left actuation tip and a right force sensing tip. Each tip is connected to a rotary arm bearing. To obtain a large gripping range, a voice coil motor (VCM) is employed to drive the left arm of the gripper through a linear guiding mechanism, which is composed of two MCPFs with $N = 2$. In addition, the sensing arm bearing is linked to two force sensor flexures which are connected in series. The sensor mechanism is similar to the compliant mechanism presented in Chapter 6.

Referring to Fig. 10.1, it is observed that the role of the actuation arm bearing is to convert the VCM driving displacement in the y-axis direction into the gripping motion of the left tip along the x-axis direction. Similarly, the role of the sensing arm bearing is to transmit the x-axis motion of the right tip into the y-axis translation of the force sensor flexures. In this way, the gripping force F_g is measured by the force sensor.

Moreover, the right tip is slightly longer than the left tip. Hence, when the gripper interacts with the environment in the y-axis direction, contact is established with the right tip first. The interaction force F_e causes a y-axis motion of the right tip, which is transferred to the force sensor flexures as well. Therefore, both the gripping and interaction forces can be detected by the force sensor.

It is notable that the gripper tip does not deliver a pure translational motion because it is guided by a rotary bearing flexure. Actually, to achieve a gripping range of several hundred to thousand micrometers, the slight rotation of gripper tips is not a critical issue, as will be verified later.

10.2.1 Actuation Issues

10.2.1.1 Driving Displacement Requirement

Referring to Fig. 10.1, the rotary bearing of the actuation arm is driven along the tangential direction by the VCM through a linear guiding flexure. The relationship between the left tip displacement x and the VCM driving displacement d_{in} is approximated by

$$\frac{x}{d_{in}} = \frac{l_4 + l_5}{l_4} \tag{10.1}$$

where l_5 is the arm length and l_4 is the driving distance between the driving point A and the rotation center O of the rotary bearing, as shown in Fig. 10.1.

Specifying a gripping displacement x of the gripper, the required driving displacement from the VCM motor can be derived from Eq. (10.1) as follows:

$$d_{in} = \frac{l_4 x}{l_4 + l_5}. \tag{10.2}$$

10.2.1.2 Driving Force Requirement

Given the driving displacement d_{in}, the required driving force F_{in} from the VCM can be determined by calculating the stiffness K_l of the left portion of the gripper, which is presented below.

The linear guiding mechanism of the VCM motor is composed of two MCPFs. Each MCPF consists of two basic modules. Hence, its spring constant can be derived from Eq. (2.11) with $N = 2$:

$$K_1 = \frac{Ebh_3^3}{l_3^3} \tag{10.3}$$

where E and b are the Young's modulus and out-of-plane thickness of the plate material, h_3 and l_3 are the in-plane width and length of the leaf flexures associated with the actuator guiding mechanism, respectively.

In addition, the rotary flexure bearing is a MCRF mechanism, which is composed of two basic modules. The torsional spring constant can be calculated in view of Eq. (8.13) with $N = 2$:

$$K_t = \frac{Ebh^3 Rr}{2l^3} \tag{10.4}$$

where the length l, and radii R and r of the leaf flexures are depicted in Fig. 8.4(a).

Assume that the tangential driving force F produces an angle θ of the rotary bearing. The torsional stiffness can be expressed as follows:

$$K_t = \frac{Fl_4}{\theta} \tag{10.5}$$

where the rotary angle θ of the bearing induces a translation d of the driving point A, which can be approximated by

$$d = l_4\theta. \tag{10.6}$$

Therefore, in view of Eqs. (10.4), (10.5), and (10.6), the equivalent linear stiffness at the driving point A can be derived as

$$K_2 = \frac{F}{d} = \frac{K_t}{l_4^2} = \frac{Ebh^3 Rr}{2l^3 l_4^2}. \tag{10.7}$$

Considering that the spring constants K_1 and K_2 are connected in parallel because both are fixed at the base, the equivalent stiffness of the left portion of the gripper can be computed as

$$K_l = K_1 + K_2 = Eb\left(\frac{h_3^3}{l_3^3} + \frac{Rrh^3}{2l^3 l_4^2}\right) \tag{10.8}$$

which is derived in view of Eqs. (10.3) and (10.7).

Hence, the required driving force from the VCM motor can be obtained as

$$F_{in} = K_l d_{in} = Eb\left(\frac{h^3}{l_3^3} + \frac{Rrh^3}{2l^3 l_4^2}\right) d_{in}. \tag{10.9}$$

10.2.1.3 Stress Constraints

To ensure that the linear guiding flexure works in the elastic condition, its displacement should stay lower than the maximum translation. The maximum translation is restricted by the yield strength, that is, the maximum allowable stress of the material.

The maximum one-sided translation of this compound parallelogram flexure can be calculated by resorting to Eq. (2.12) with $N = 2$:

$$d_{\mathrm{max1}} = \frac{2\sigma_y l_3^2}{3Eh_3} \tag{10.10}$$

where σ_y is the yield strength of the material.

Thus, the driving displacement should meet the following condition:

$$d_{\mathrm{in}} \leq d_{\mathrm{max1}} = \frac{2\sigma_y l_3^2}{3Eh_3} \tag{10.11}$$

to guarantee that no plastic deformation occurs for the linear guiding flexure.

In addition, the maximum one-sided rotary angle of the rotary flexure can be derived in view of Eq. (8.18) with $N = 2$:

$$\theta_{\mathrm{max}} = \frac{2\sigma_y l^2 (R + r)}{3ERrh}. \tag{10.12}$$

To guarantee that the rotary bearing operates with elastic deformations, the driving displacement should also satisfy the following condition:

$$d_{\mathrm{in}} \leq d_{\mathrm{max2}} = l_4 \theta_{\mathrm{max}} = \frac{2\sigma_y l^2 l_4 (R + r)}{3ERrh}. \tag{10.13}$$

10.2.2 Position and Force Sensing Issues

The position of the left tip can be measured by a strain gauge-based displacement sensor, which is glued onto the surface of the leaf flexures associated with the actuator guiding mechanism.

In addition, the gripping force F_g for the object or the interaction force F_e with the environment exerted by the right tip is transmitted to the force sensor flexures in the right portion of the gripper structure. Specifically, the force sensor #1 is composed of four fixed–guided flexures, which experience an identical deformation due to having the same dimensions. The force sensor #2 consists of two MCPFs with $N = 2$. Attaching strain gauges on these flexures, the force experienced by the right tip can be measured. The sensor specifications are designed as follows.

10.2.2.1 Sensor Sensitivity Design

According to Hooke's law, the force experienced by the two sensor flexures can be calculated as

$$F_1 = K_{s1} y_1 \tag{10.14}$$

$$F_2 = K_{s2} y_2 \tag{10.15}$$

with

$$K_{s1} = \frac{Ebh_1^3}{l_1^3} \tag{10.16}$$

$$K_{s2} = \frac{Ebh_2^3}{l_2^3} \tag{10.17}$$

where l_1 and h_1 are the in-plane length and width of the flexures associated with sensor #1, and l_2 and h_2 are the in-plane length and width of the flexures related to sensor #2, respectively. In addition, y_1 and y_2 are the displacements of the two sensors.

When the two sensors are connected in series, they experience the same force (i.e., $F_1 = F_2$). Then, it is deduced from Eqs. (10.14) and (10.15) that the smaller the stiffness is, the larger the displacement will be. A larger displacement induces a larger strain of the strain gauge, producing a higher magnitude of the output signal. As a result, a higher sensing sensitivity is obtained.

The sensing scheme is shown in Fig. 10.2. In this work, sensor flexure #1 is designed to have a lower stiffness than that of sensor flexure #2. Therefore, force sensor #1 exhibits a higher sensitivity than sensor #2. In addition, force sensor #1 is designed to have a smaller measurement range, which is constrained by a mechanical stopper. After experiencing a certain magnitude of applied force, the translation of sensor flexure #1 is stopped by stopper #1, which is mounted between the structures of sensors #1 and #2, as shown in Fig. 10.2. Additionally, force sensor #2 has a lower sensing sensitivity in a larger measuring range, which is constrained by stopper #2. By realizing a two-stage sensing scheme design, sensor #1 is suitable to detect

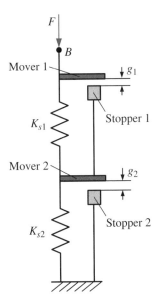

Figure 10.2 Schematic of two-stage force sensing scheme of the gripper tip.

initial contact of the gripper with the grasped object or environment with a fine resolution, and sensor #2 is able to measure the force after contact in a larger measuring range.

Without loss of generality, the output signal of the strain gauge is conditioned using a quarter-bridge circuit as shown in Fig. 6.4. The output voltage of the circuit is approximated by

$$V = \frac{V_s}{4R} \cdot \Delta R \tag{10.18}$$

where V_s is the excitation voltage, R is the nominal resistance, and ΔR is the resistance change value of the strain gauge.

The relationships between the displacements (y_1, y_2) of the two sensors and the circuit output voltages (V_1, V_2) can be established as follows [12]:

$$y_1 = \frac{8l_1^2 V_1}{3h_1 S_1 V_s} \tag{10.19}$$

$$y_2 = \frac{16l_2^2 V_2}{3h_2 S_2 V_s} \tag{10.20}$$

where S_1 and S_2 are the gauge factors of strain gauges 1 and 2, respectively. In addition, V_s is the source voltage of the Wheatstone bridge circuit.

In view of Eqs. (10.14)–(10.20) and the relation $F_1 = F_2$, the ratio of the sensitivities of the two strain gauges can be generated as

$$\eta = \frac{\frac{V_1}{F_1}}{\frac{V_2}{F_2}} = \frac{V_1}{V_2} = \frac{2S_1 h_2^2 l_1}{S_2 h_1^2 l_2}. \tag{10.21}$$

10.2.2.2 Measuring Range Design

From the foregoing analysis, it is observed that in the smaller force measuring range $[0, F1]$, the deformation is experienced by both sensor flexures. Meanwhile, in the larger force range $[F1, F2]$, the deformation of the sensing flexures is attributed to sensor #2 alone.

Considering the serial connection of the equivalent linear springs, the overall stiffness K_{R1} of the sensing system in the smaller range $[0, F1]$ can be derived as

$$K_{R1} = \frac{1}{\frac{1}{K_{s1}} + \frac{1}{K_{s2}}} = \frac{Ebh_1^3 h_2^3}{h_1^3 l_2^3 + h_2^3 l_1^3}. \tag{10.22}$$

Once a particular displacement $S1$ of the sensing point B is caused by the corresponding force $F1$, mover #1 translates a displacement g_1 relative to stopper #1 and contacts it. At this moment, the driving force can be expressed as follows:

$$F1 = K_{R1} \cdot S1 = K_{s1} \cdot g_1 \tag{10.23}$$

where the clearance g_1 is shown in Fig. 10.2. The clearance can be adjusted by tuning the clearance adjuster as shown in Fig. 10.3. In addition, the displacement stroke $S1$ can be determined from Eq. (10.23).

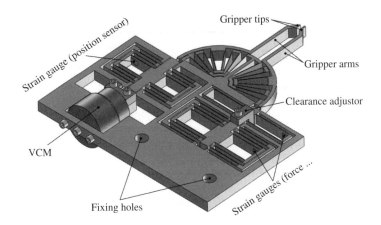

Figure 10.3   A CAD model of the designed gripper driven by a voice coil motor.

Moreover, in view of Eqs. (10.16) and (10.23), the force range #1 can be derived as

$$F1 = \frac{Ebh_1^3 g_1}{l_1^3}.$$ (10.24)

Afterwards, if the force is continually applied to the sensing element, the sensor flexure #1 will not be deformed any more because it has already been restricted by stopper #1. Under such a situation, the force condition of the flexures associated with force sensor #1 will not change any more. The force cannot be measured by force sensor #1 because it is saturated. Instead, the force is only measured by force sensor #2 because its flexures suffer from further deformations. Thus, the force can be considered to be applied to sensor flexure #2 alone. As a result, the overall stiffness in the larger measuring range is

$$K_{R2} = K_{s2}.$$ (10.25)

Next, when mover #2 contacts stopper #2, force sensor #2 is saturated, and no more flexure elements will be deformed. The extreme of the force range #2 can be determined by

$$F2 = K_{R2} \cdot g_2.$$ (10.26)

Hence, the extreme of the measuring range #2 is derived as follows:

$$F2 = \frac{Ebh_2^3 g_2}{l_2^3}.$$ (10.27)

It is notable that the clearances g_1 and g_2 are much smaller than the maximum displacements of the sensor flexures #1 and #2, respectively. In addition, the induced displacement of the sensing point B does not cause plastic deformation of the sensing-arm rotary bearing.

Table 10.1   Main parameters of a compliant gripper

Parameter	Symbol	Value	Unit
Outer radius of MCRF	R	26.5	mm
Inner radius of MCRF	r	10.0	mm
Flexure length of MCRF	l	16.5	mm
Flexure width of MCRF	h	0.3	mm
Flexure length of force sensor #1	l_1	25.0	mm
Flexure width of force sensor #1	h_1	0.3	mm
Flexure length of force sensor #2	l_2	20.0	mm
Flexure width of force sensor #2	h_2	0.3	mm
Flexure length of actuator guiding mechanism	l_3	15.0	mm
Flexure width of actuator guiding mechanism	h_3	0.3	mm
Driving distance	l_4	30.0	mm
Arm length of the gripper	l_5	35.0	mm
Out-of-plane thickness of material	b	6.35	mm
Clearance between mover #1 and stopper #1	g_1	0.2	mm
Clearance between mover #2 and stopper #2	g_2	0.5	mm

10.3   Performance Evaluation with FEA Simulation

As a case study, a gripper is designed using a piece of Al 7075 alloy material. A CAD model of the gripper is illustrated in Fig. 10.3 and the main structural parameters are shown in Table 10.1. The design objective is to achieve a large gripping range of [0, 4.40 mm] between the two gripper tips.

10.3.1   Analytical Model Results

To generate a gripping displacement of ± 2.20 mm for the left tip, the required input displacement from the VCM motor is calculated by Eq. (10.2) as ± 1.0 mm, which needs an actuation force of 4.11 N. The driving displacement is much lower than the one-sided allowable translations of $d_{max1} = 3.5$ mm and $d_{max2} = 17.5$ mm, as evaluated by Eqs. (10.11) and (10.13), respectively.

To implement the force sensor, a piezoresistive and a metal strain gauge are attached to sensor flexures #1 and #2, respectively. Assume that the gauge factors for the two strain gauges are $S_1 = 130$ and $S_2 = 2.1$, respectively. Using Eq. (10.21), the ratio of the two sensitivities is calculated as $\eta = 154.76$. In addition, based on Eqs. (10.24) and (10.27), the two force measuring ranges are evaluated as [0, 0.1573 N] and [0, 0.7683 N], respectively.

10.3.2   FEA Simulation Results

To verify the performance of the stage, FEA simulations are carried out using the ANSYS software package.

10.3.2.1 Performance of Actuation Arm

First, the statics performance of the designed gripper is evaluated using static structural FEA. The simulations are conducted by applying an input force of 1 N to the input end of the left arm of the gripper. The induced overall deformation of the gripper is shown in Fig. 10.4(a).

To produce a gripping displacement of ± 2.20 mm, simulation results indicate that the required driving displacement is ± 1.07 mm and the required maximum driving force is 5.71 N from the motor. Taking the FEA simulation results as benchmark, it is observed that the analytical models underestimate the required driving displacement and force by 6.5% and 28.0%, respectively. The discrepancies are mainly induced by the assumption used in analytical models, where only the bending deformations of the flexures associated with the linear and rotary bearings are considered. More accurate results can be produced by establishing analytical models in consideration of the deformations of the entire compliant mechanism.

In addition, FEA simulation reveals that the maximum stress occurs around the flexure at the driving point A. To produce the required gripping range of ± 2.20 mm for the left tip, the induced maximum stress is 34.6 MPa, which is far less than the yield strength of 503 MPa for the material. Hence, the elastic deformation of the material is well guaranteed by the designed parameters shown in Table 10.1.

To create the full gripping range of ± 2.20 mm, FEA simulation reveals that the induced rotational angle of the gripper tip is $\pm 1.76°$. To deliver a smaller gripping range of ± 0.50 mm, the rotational angle of the gripper tip is only $\pm 0.40°$. Thus, the slight rotation of the gripper tip can be neglected as it will not have much influence on the grasping process.

10.3.2.2 Performance of Sensing Arm

When a gripping force $F_g = 1$ N and an environment interaction force $F_e = 1$ N are applied at the right tip, the FEA simulation results of the deformations for the gripper are shown in Fig. 10.4(b) and (c), respectively.

It is known that the output voltage V_o of the strain gauge is related to the deflection of each leaf flexure μ as follows [12]:

$$V_o = \frac{3SV_sK_fl\mu}{4Ebh^2} \tag{10.28}$$

where l and h represent the length and width of leaf flexures, S is the gauge factor of the strain gauge, b is the thickness of the material, K_f is the linear stiffness of each leaf flexure, and V_s is the source voltage of the quarter-bridge circuit, respectively.

Hence, the ratio of the sensitivities of the two strain gauges can be derived as

$$\eta_{FEA} = \frac{V_1/F_1}{V_2/F_2} = \frac{V_1}{V_2} = \frac{S_1l_1h_2^2K_{f1}\mu_1}{S_2l_2h_1^2K_{f2}\mu_2}. \tag{10.29}$$

To assess the stiffness of the two sensor flexures, an FEA simulation is carried out by applying a force to the two sensor flexure mechanisms and the simulation result of the deformation is shown in Fig. 10.5. By extracting the deflections μ_1 and μ_2 and the stiffnesses K_{f1} and K_{f2} of the flexures associated with the two sensor mechanisms, the sensitivity ratio is calculated as

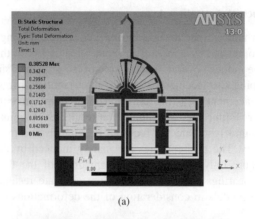

(a)

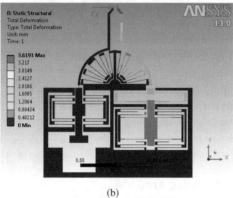

(b)

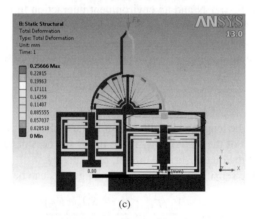

(c)

Figure 10.4 Static structural FEA simulation results of the gripper: (a) actuation force is applied; (b) gripping force is applied; (c) interaction force is applied.

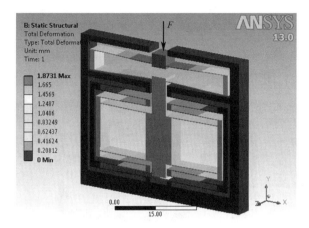

Figure 10.5 FEA simulation results of the sensor flexure with a force applied.

Table 10.2 Gripper performances evaluated by analytical models and FEA simulations

Performance	Model result	FEA simulation result	Model error (%)
d_{in} (mm)	±1.00	±1.07	6.5
F_{in} (N)	4.11	5.71	28.0
$F1$ (N)	0.1573	0.1619	2.8
$F2$ (N)	0.7683	0.7847	2.1
η	154.76	183.18	15.5

$\eta_{FEA} = 183.18$. Compared with the FEA results, the analytical model underestimates the sensitivity ratio by 15.5%. The deviation arises from the assumption used in the analytical model, which only considers the bending deformations of the leaf flexures. The modeling error can be reduced by taking into account all kinds of deformations of the mechanism.

In addition, using the stiffnesses K_{s1} and K_{s2} of the two sensor flexures, which are generated by the aforementioned FEA simulations, the force sensing ranges are determined as $F1 = 0.1619$ N and $F2 = 0.7847$ N, respectively. Compared with the simulation results, the analytical models predict the two ranges with small errors of 2.8% and 2.1%, respectively. For a clear comparison, the results of the analytical models and FEA simulations are summarized in Table 10.2.

10.3.2.3 Structure Dynamics Performance

To evaluate the dynamics performance of the gripper structure, a modal analysis simulation is performed. FEA simulation results of the first two resonant modes are shown in Fig. 10.6.

It is observed that the first mode is induced by the motion of the sensing arm with a resonant frequency at 66.23 Hz, and the second resonant mode at 96.96 Hz is mainly attributed to

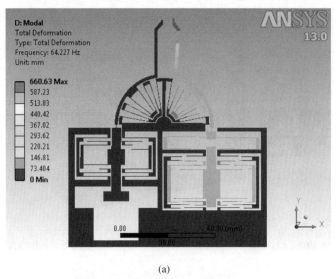

(a)

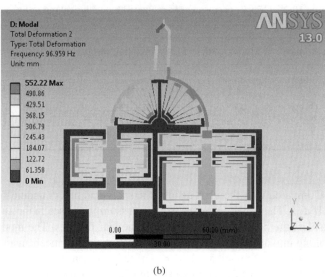

(b)

Figure 10.6 The first (a) and second (b) resonant modes of the gripper structure.

the translation of the actuation arm. The simulation results give an intuitive prediction of the resonant mode shapes of the gripper structure.

10.3.2.4 Fatigue-Lifecycle Analysis

Fatigue analysis is important in the mechanical design stage of the gripper. To produce a desired gripping range, the fatigue life of the gripper structure is predicted through FEA

Table 10.3 FEA simulation results of fatigue lifecycles producing different gripping ranges

Gripping range (mm)	Driving displacement (mm)	Lifecycle
±0.50	±0.243	1.0000×10^8
±1.00	±0.486	1.0889×10^7
±2.00	±0.973	3.2983×10^4
±2.20	±1.070	1.0057×10^4

simulation. Specifically, by applying a fully reversed load corresponding to the desired gripping range to the driving end of the left arm, stress life analysis is carried out.

By specifying different gripping ranges, the simulation results of the lifecycles are shown in Table 10.3. To produce the maximum gripping range of ±2.20 mm, FEA simulation predicts that the maximum number of lifecycles is about 10^4 before fatigue failure of the structure. In addition, the number of lifecycles increases significantly with a reduction in required gripping range. A gripping range of less than 1 mm will lead to an approximately infinite life of the gripper structure.

10.4 Prototype Development and Calibration

10.4.1 Prototype Development

The prototype gripper is shown in Fig. 10.7. The gripper is fabricated from a piece of Al 7075 plate using the wire-electrical discharge machining (EDM) process. To drive the gripper, a VCM (model: NCC04-10-005-1A, from H2W Techniques, Inc.) is chosen, which is able to

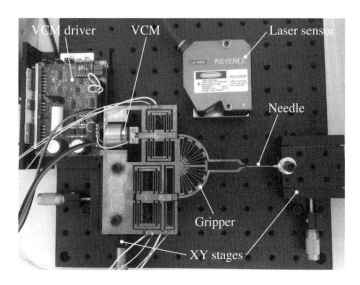

Figure 10.7 Experimental setup of a prototype gripper system.

provide the maximum driving force of 29.2 N with a stroke of 10.2 mm. It is actuated by the NI-9263 analogy output module (from National Instruments Corp.) through a VCM driver. For measuring the gripper's left tip position, a metal strain gauge (model: SGD-3/350-LY13, from Omega Engineering Ltd.) is employed to construct a quarter-bridge circuit. The strain gauge has a gauge factor of 2.1 and a nominal resistance of 350 Ω. To measure the gripping and interaction forces exerted on the right tip of the gripper, force sensors #1 and #2 are realized using a piezoresistive strain gauge (model: HU-101A, from Bengbu Heli Sensing System Engineering Co., Ltd.) and a metal strain gauge (model: SGD-3/350-LY13, from Omega Engineering Ltd.), respectively. The piezoresistive strain gauge has dimensions of 6 mm × 4 mm, a nominal gauge factor of 130, and a nominal resistance of 350 Ω. It is more sensitive than the metal strain gauge.

In addition, an NI cRIO-9022 real-time (RT) controller combined with a cRIO-9118 reconfigurable chassis (from National Instruments Corp.) is adopted to implement the control algorithm. The modules of NI-9263 analog output and NI-9237 simultaneous bridge input are adopted for the creation of analog excitation signals and the acquisition of strain-gauge sensor signals, respectively. The NI-9237 input module provides a 24-bit analog-to-digital conversion resolution. The NI cRIO-9118 chassis contains a FPGA core, and the associated cRIO-9022 RT controller communicates with a personal computer via the Ethernet port.

For the calibration of the strain-gauge position sensor, the tip position of the gripper is also measured by a laser displacement sensor (model: LK-H055, from Keyence Corporation), which provides a sub-micrometer resolution within a 20-mm measuring range. Additionally, the prototype is programmed using NI LabVIEW software to realize real-time control for the gripper system.

10.4.2   Calibration of Position Sensor

The purpose of the position sensor calibration is to determine the sensitivity of the strain gauge so as to convert its output voltage into a position value. The sensitivity can be derived by comparing the strain-gauge output voltage with the reference position values. It has been shown that the sensitivity can be calibrated by using different input signals, such as sinusoidal and square waves [13]. In this work, sinusoidal signals are adopted for the calibration of the position sensor.

Specifically, a 0.2-Hz sine wave with 1.5-V amplitude is employed to drive the VCM. The output position of the left tip is measured by the strain gauge and the laser sensor, as shown in Fig. 10.8(a) and (b), respectively. It is observed that the left tip exhibits a displacement range of [−2.18 mm, 2.22 mm], which satisfies the design requirement. The root-mean-square (RMS) values of the strain-gauge output voltage and laser sensor output position are 2.7399×10^{-4} V and 1.4603×10^{3} μm, respectively. Hence, a position sensitivity of 1.8763×10^{-7} V/μm is derived, which is used to convert the strain-gauge output voltage into a position value.

With the calibrated strain-gauge position sensor, the relationship between the applied voltage and the output displacement of the gripper tip is shown in Fig. 10.8(c). It is observed that there is no clear hysteresis in the actuation of the left gripper tip. Actually, the gripper arm can also be driven using a higher-voltage input, which drives the left tip to contact the right tip in the closing operation process.

With a zero-voltage input, the noise of the strain-gauge position sensor is recorded as shown in Fig. 10.8(d). It is seen that the noise closely follows a normal distribution with standard

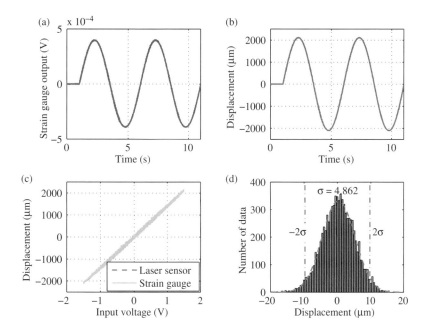

Figure 10.8 Calibration results of strain-gauge position sensor: (a) output voltage of the strain gauge; (b) output displacement of the laser sensor; (c) input–output relations obtained by the calibrated strain-gauge sensor and laser sensor; (d) histogram of strain-gauge position sensor noise.

deviation (σ) of 4.862 µm. By adopting 2σ as the resolution, the position sensor provides a resolution of 9.724 µm. That is, the minimum detectable step of the displacement of the left gripper tip is 9.724 µm.

10.4.3   Calibration of Force Sensor

To calibrate the strain-gauge force sensor, a hung weight of 25 mN is tied to the right tip of the gripper using a nylon wire. Initially, the weight is supported without exerting a force on the gripper tip. By removing the support of the weight suddenly, a force is applied to the right tip, which induces the voltage outputs of the two strain gauges, as shown in Fig. 10.9(a) and (b), respectively. At steady state, the two strain gauges produce output voltages of 8.3467×10^{-4} and 4.1222×10^{-6} V, respectively. The sensitivity values of force sensors #1 and #2 are derived as 3.339×10^{-5} and 1.649×10^{-7} V/mN, respectively. This leads to a sensitivity ratio of $\eta = 202.48$. Compared with the simulation results, the experimental result of the sensitivity ratio is 10.5% higher. The discrepancy originates mainly from the manufacturing tolerance of the gripper structure and the difference between the nominal and actual values of the gauge factors.

The force sensitivities are used to convert the voltages into force values, as shown in Fig. 10.9(c) and (d), respectively. In addition, Fig. 10.9(e) and (f) exhibits the noise histograms of two force sensors without a force exerted. The 2σ resolutions of the fine and coarse force sensors are calculated as 0.314 and 20.512 mN, respectively. The resolution ratio

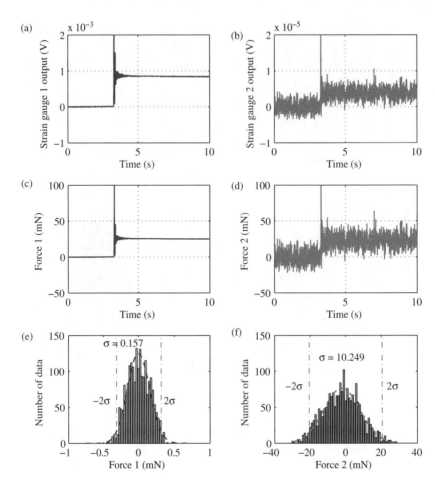

Figure 10.9 Calibration results of strain-gauge force sensor: (a) output voltage of strain gauge #1; (b) output voltage of strain gauge #2; (c) output force of calibrated strain-gauge force sensor #1; (d) output force of calibrated strain-gauge force sensor #2; (e) noise histogram of force sensor #1; (f) noise histogram of force sensor #2.

of the two sensors can be derived as 1/65.3. Therefore, the resolution in the smaller range has been improved 65.3 times compared with that in the larger force range.

Actually, the resolution is determined by the analog-to-digital conversion resolution of the hardware (i.e., 24-bit NI-9237 in this work) and the electric noise. It is observed that the resolution ratio is different from the sensitivity ratio of the two force sensors. The main reason lies in the different noise levels of the employed piezoresistive and metal strain gauges.

10.4.4   Verification of Force Sensor

The two force sensors are calibrated using a known weight as conducted above. To verify the strain sensors, the sensor outputs are compared with the forces measured using a commercially

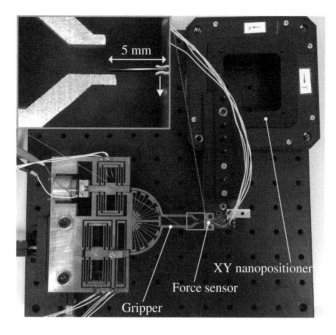

Figure 10.10 Experimental setup for the verification of strain-gauge force sensors.

available force sensor (model: AE801, from Kronex Technologies Corp.). The experimental setup is shown in Fig. 10.10, where the sensor AE801 is mounted on an XY nanopositioner (model: P-734.2CL, from Physik Instrumente Co., Ltd.) and the gripper is fixed on a manual XY stage (model: M-460A-XY, from Newport Corp.). Before the experiment, the gripper is positioned such that the sensing tip is adjacent to the sensor AE801. Then, the sensor AE801 is carried by the nanopositioner to contact the sensing tip of the gripper. During the translation of 100 μm displacement, both force sensor AE801 and strain-gauge sensors produce force readings.

Figure 10.11 shows the output signals of the force sensors. It is observed that when the nanopositioner translates consecutive steps of 4-μm size in forward and backward directions, force sensor #1 delivers a similar output to the commercial force sensor AE801, as shown in Fig. 10.11(a). The average discrepancy between the readings of the two sensors is derived as 0.061 mN, which verifies the accuracy of force sensor #1. However, due to the small magnitude of the induced force, the noisier force sensor #2 cannot discriminate the variation steps of the force readings, as depicted in Fig. 10.11(b). Even so, force sensor #2 reflects the ascending and descending trends of the force and exhibits the minimum and maximum force values (i.e., 0 and 15.2 mN) during the translation of the nanopositioner.

10.4.5 Consistency Testing of the Sensors

In order to test the consistency of the strain-gauge position and force sensors, the gripper arm is driven using a constant input signal of 2 V to grasp a steel needle. The output signals of the

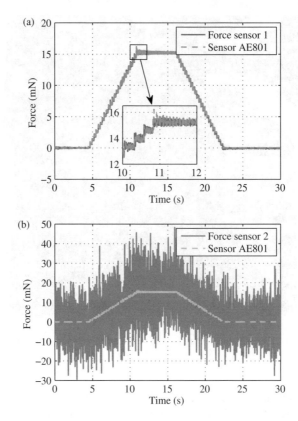

Figure 10.11   Sensor readings verified using a commercial force sensor: (a) comparison of force sensor #1 and sensor AE801 outputs; (b) comparison of force sensor #2 and sensor AE801 outputs.

position sensor and two force sensors over 10 minutes are shown in Fig. 10.12. It is observed that all three sensors produce consistent results over a long period of time. Compared with the other two sensors under the same environmental conditions (room temperature), the output of force sensor #1 exhibits a slight fluctuation around 83.9 mN, as shown in Fig. 10.12(c). The reason lies in the fact that the piezoresistive strain gauge is more sensitive to temperature change. To remedy this issue, a signal conditioning circuit using a half or full bridge can be employed.

10.5   Performance Testing Results

Using the calibrated position and force sensors, the gripping and interaction-detection performances of the gripper are tested by experimental studies.

10.5.1   Testing of Gripping Sensing Performance

The gripping performance of the gripper with integrated position and force sensors is tested by gripping and releasing a microneedle with a diameter of 300 μm, as shown in Fig. 10.7. The

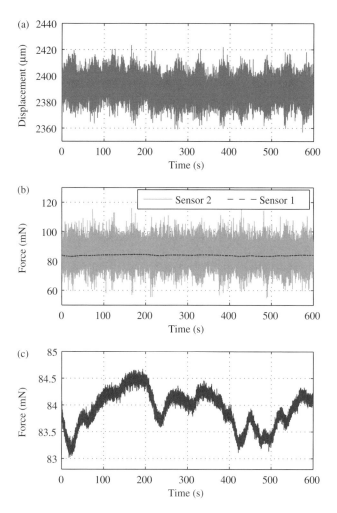

Figure 10.12 Consistency testing results of the sensors: (a) strain-gauge position sensor; (b) two strain-gauge force sensors; (c) force sensor #1.

needle is fixed between the two tips of the gripper. The initial distances between the needle and the left and right tips are d_l and d_r, respectively. Moreover, the needle is closer to the right tip, with $d_r < d_l$.

To conduct the gripping task, a voltage as shown in Fig. 10.13(a) is used to drive the VCM, and the displacement of the left tip is measured by the strain-gauge position sensor, as shown in Fig. 10.13(b). The gripping force is measured by the two strain-gauge force sensors, as plotted in Fig. 10.13(c) and (d), respectively. It is observed that force sensor #1 saturates on the boundary of [0, 138.3 mN], which represents the smaller force measurement range of the gripper. Within this smaller range, the force can be measured by both sensors, while sensor #2 produces a lower resolution than sensor #1. In the force range larger than 138.3 mN, the gripping force is measured by force sensor #2 alone.

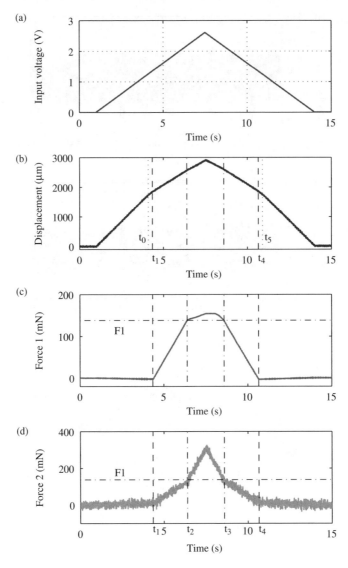

Figure 10.13   Gripping performance testing results of the gripper: (a) input voltage; (b) displacement of the left tip; (c) gripping force measured by force sensor #1; (d) gripping force measured by force sensor #2.

Referring to Fig. 10.13(d), it is observed that the right tip of the gripper contacts the needle at time t_1. During the time interval between t_1 and t_2, the gripping force is measured by both force sensors #1 and #2, although the former provides a much higher resolution than the latter. Between t_2 and t_3, sensor #1 is saturated around 138.3 mN, as shown in Fig. 10.13(c), and the gripping force is measured by sensor #2. In this time interval, the varying output of sensor #1 may be caused by the movement of the leads attached to strain gauge #1.

In addition, the larger force sensing range is tested by applying a sufficiently large force to the right arm. Experimental results reveal that sensor #2 saturated around 815.2 mN. Hence, sensor #2 provides a force measuring range of [0, 815.2 mN]. The discrepancy between the simulation and experimental results of the two force ranges is mainly caused by the nominal and actual values of the clearances g_1 and g_2. Actually, the force ranges can be tuned by adjusting these two clearances.

10.5.2   Testing of Horizontal Interaction Detection

The performance of the force sensor for horizontal contact detection of the gripper with the environment is tested by applying a force to the right tip. In particular, the force is exerted by the microneedle, which is fixed on an XY stage driven by two micrometers. For this testing, the relative position of the needle with respect to the gripper tip is adjusted under a microscope by tuning the two micrometers. Then, the contact force F_e is applied to the right tip by manually tuning the micrometers to mimic the interaction with an environment.

The contact force is measured by the two force sensors, as shown in Fig. 10.14. It is observed that both sensors can detect the interaction force F_e which is applied to the right tip. Different from the gripping force measurement, sensor #1 produces a negative force signal. The different signs of the sensor #1 outputs can be explained by examining the deformation results, as shown in Fig. 10.4(b) and (c).

Specifically, in the gripping process, the flexures associated with force sensor #1 suffer from a common force applied at point B along the y-axis. Meanwhile, during the interaction of the right tip with the environment, point B undergoes an x-axis transverse translation in addition to the translation in the y-axis direction. The x-axis translation is induced owing to a low transverse stiffness of sensor flexure #1. As a consequence, the four leaf flexures exhibit a deformation which is different from that in the gripping process. Hence, a negative strain is

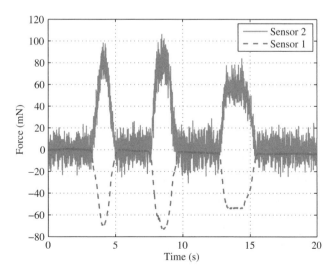

Figure 10.14   Horizontal interaction force detected by force sensors #1 and #2.

induced, which leads to a negative output force signal during horizontal interaction with the environment.

10.5.3   Testing of Vertical Interaction Detection

The vertical contact detection is demonstrated by mounting the gripper on a micromanipulator. Figure 10.15 shows the experimental setup used. The gripper is mounted on an XYZ micromanipulator (model: MP-285, from Sutter Instrument Inc.) with an inclined angle with respect to the horizonal plane. When the gripper is lowered, the right tip makes contact with the plane first because the right tip is longer than the left one. Hence, the vertical contact can be detected by the sensing arm. In this work, the gripper is lowered to contact the weight-supporting stage of a microbalance (model: HZK-FA110, from Huazhi Scientific Instrument Co., Ltd.). The microbalance offers a mass resolution of 0.1 mg, which is equivalent to a force resolution of 0.98 μN.

When the gripper is translated 5.5 mm in the $-z$ direction to contact the microbalance, the outputs of the two force sensors are recorded as shown in Fig. 10.16. It is observed that the gripper keeps still before t_1. Starting from t_1, the gripper is lowered at a constant velocity of 1.7 mm/s. The gripper contacts the microbalance at t_2 and stops at t_3. At the steady state of the contact, the microbalance displays a mass of 34.4 g, which is equivalent to 337.361 mN.

At the start (t_1) and stop (t_3) moments of the translation, force sensor #1 exhibits a clear oscillation in the output signal. The oscillation arises from the vibration of the flexible sensor structure, which is induced by the inertia of the gripper. Due to a relatively high level of noise of force sensor #2, the output of sensor #2 is not influenced by the inertia effect. It is notable that the vibration is introduced by the micromanipulator, which starts translation with a constant velocity of 1.7 mm/s at t_1 and stops suddenly at t_3. The vibration can be mitigated by a proper trajectory planning of the micromanipulator motion, e.g., starting and stopping with zero acceleration.

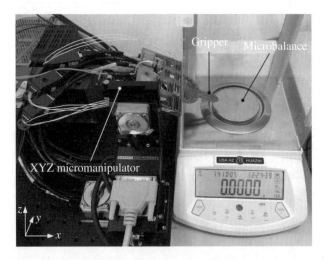

Figure 10.15   Experimental setup for vertical interaction detection of the gripper.

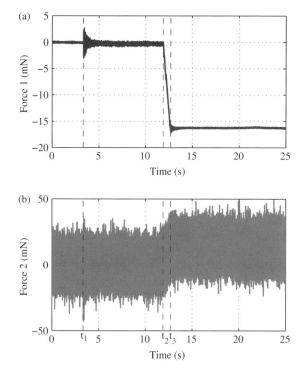

Figure 10.16 Vertical interaction force detected by force sensors #1 (a) and #2 (b).

By examining the sensor outputs at the steady state, it is found that sensors #1 and #2 predict vertical contact forces of −16.226 mN and 12.809 mN, respectively. The different signs of the two sensor outputs are caused by the different deformation directions of the sensing flexures when vertical force is applied. Compared with the microbalance reading, the outputs of sensors #1 and #2 are about 20 and 25 times lower. The reason is that the two sensors are calibrated by applying a gripping force in the gripper plane. When the gripper suffers from a force vertical to the gripper plane, the sensitivities of the two force sensors are much smaller. Hence, although the two force sensors can detect the vertical interaction force, their sensitivities are over 20 times smaller than those of gripping force and horizontal contact force sensing.

10.5.4 Testing of Dynamics Performance

Generally, the dynamic characteristics of a system can be evaluated by means of impact response, step response, and frequency response. Regarding the actuation arm of the gripper, FEA simulation predicts a natural frequency lower than 100 Hz. To excite the resonance of the actuation arm, the frequency response method is employed. In particular, by fixing the gripper on the base, the swept-sine waves with an amplitude of 0.03 V and frequency of interest ranging from 1 to 500 Hz are used to drive the VCM. The displacement response is measured by the strain-gauge position sensor. Spectral analysis is then conducted to derive the frequency responses of the gripper, as shown in Fig. 10.17. A resonant peak at 42.66 Hz is observed, which is lower than the frequency predicted by FEA simulation. The discrepancy

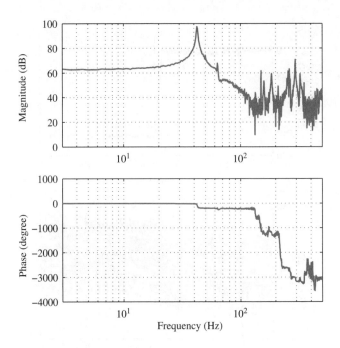

Figure 10.17   Bode plots of the frequency response of the actuation arm of the gripper.

between the experimental and simulation results of the resonant frequency mainly arises from the manufacturing tolerance and the added mass of the VCM, which is not considered in FEA simulation.

Next, the dynamics performance of the sensing arm of the gripper has been tested. In this work, the step response approach is employed due to its efficiency and easy implementation. Initially, a mass of 8.3767 g is hung by a nylon wire at the gripper tip of the sensing arm. Due to the gravity (9.807 m/s^2) effect, the two strain-gauge force sensors produce readings of 82.15 mN at steady state. When the wire is cut, the force imposed on the sensing arm is removed suddenly, which induces a negative step response of the right tip. The responses captured by force sensors #1 and #2 are shown in Fig. 10.18(a) and (b), respectively. The magnitudes of frequency responses are obtained by fast Fourier transform (FFT) and shown in Fig. 10.18(c) and (d), respectively. Owing to a higher resolution, force sensor #1 produces a smoother result than sensor #2. In addition, both force sensors predict the first two resonant frequencies around 45.23 and 68.36 Hz. Compared with FEA simulation results, the experimental result of 45.23 Hz for the first natural frequency is 31.7% lower. The discrepancy is mainly caused by the manufacturing tolerance and the mass of the clearance adjustor, which is not considered in the FEA simulation.

10.5.5   Applications to Pick–Transport–Place in Assembly

To demonstrate the performance of the developed gripper in assembly applications, the gripper is mounted on an XYZ micromanipulator. The experimental setup is the same as shown in

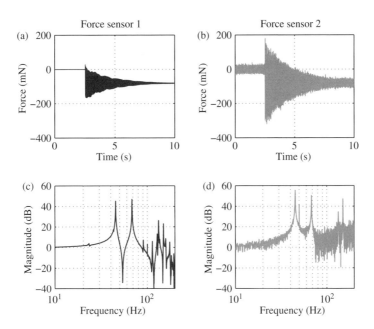

Figure 10.18 (a), (b) Negative step responses of the sensing arm of the gripper obtained by force sensors #1 and #2; (c), (d) magnitudes of frequency responses obtained by force sensors #1 and #2.

Fig. 10.15. A trajectory is planned to pick, transport, and place a small (M2) screw onto a microbalance. The experimental results are shown in Fig. 10.19.

Referring to Fig. 10.19(c), in time slot 1, the gripper is translated 11 mm in the $-z$ direction at a constant velocity of 2.9 mm/s. Then, the gripper is driven to grasp the screw in time slot 2. In the subsequent time slot 3, the gripper is translated 11 mm upward along the z-axis to pick up the screw. Afterwards, the gripper is translated by 25 mm in the x-axis direction at a velocity of 2.9 mm/s in time slot 4 to transport the screw above the microbalance. In time slot 5, the gripper is lowered by 2 mm along the $-z$ direction and the grasped screw contacts the microbalance, which causes a reading of the microbalance larger than the screw weight. Because the sensitivities of the force sensors are more than 20 times lower than their normal values, the influence of the vertical force on the sensor outputs is not obvious. In time slot 6, the screw is released on the weight-supporting stage of the microbalance. The gripper is then translated 2 mm upward along the z-axis in time slot 7, and returns to the home position by translating 25 mm along the $-x$ direction during time slot 8.

As mentioned before, due to the inertia effect of the gripper, the output of force sensor #1 exhibits an oscillation phenomenon when the XYZ micromanipulator starts and stops moving at a constant velocity, as shown in Fig. 10.19(c). In contrast, force sensor #2 is not influenced by inertia due to a larger noise level, as depicted in Fig. 10.19(d). Moreover, Fig. 10.19(b) reveals that the position sensor output is not affected by the inertia effect either.

10.5.6 Further Discussion

Driven by a constant-slope voltage signal as shown in Fig. 10.13(a), the displacement curve of the left tip of the gripper reveals different slopes during the gripping and releasing operation, as

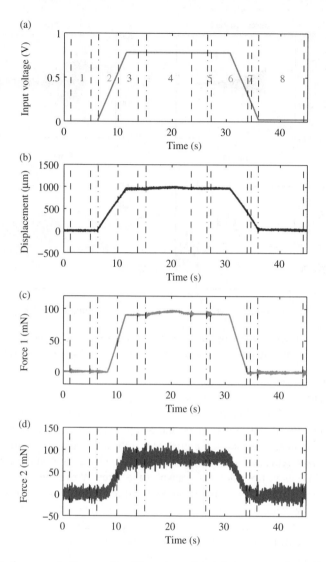

Figure 10.19 Performance testing results of the gripper for pick–transport–place of a screw: (a) input voltage; (b) displacement of the left tip; (c) gripping force measured by force sensor #1; (d) gripping force measured by force sensor #2.

shown in Fig. 10.13(b). It is found that the displacement increases with initial rising of the driving voltage. At moment t_1, the right tip contacts the microneedle, as indicated in Fig. 10.13(c) and (d). However, the slope of the displacement curve for the left tip begins to change at t_0 before t_1. This phenomenon reveals that the left tip contacts the needle at t_0 first, then the left tip pushes the needle to contact the right tip at t_1. Hence, by comparing the left tip displacement at t_0 and t_1, the initial clearance between the needle and the right tip can be derived as $d_r = 110$ μm. In addition, the initial distance between the needle and the left tip can be determined as the displacement at t_0 (i.e., $d_l = 1810$ μm). Furthermore, the initial distance between the two tips of the gripper can be calculated as the sum of the two clearances d_l and d_r and the needle diameter of 2220 μm. In view of this initial distance and the gripping displacement of the left tip, a gripping range around [0, 4.40 mm] can be derived, which is consistent with the design objective.

In addition, Fig. 10.13(d) exhibits different slopes of the force curves for sensor #2. Actually, this phenomenon reveals different stiffnesses of the gripper during the two-stage force sensing in the gripping process. The force–displacement relationship is illustrated in Fig. 10.20. It is observed that the gripping and releasing curves are overlapped with each other, which means that no hysteresis effect exists. Between t_1 and t_2 (i.e., in the smaller force range) the gripper stiffness can be derived as the slope $K_1 = 0.1869$ mN/μm of the best linear fit of the curves produced by both force sensors. Meanwhile, between t_2 and the peak time of the force (i.e., within the larger force range) the stiffness value is predicted by sensor #2 as $K_2 = 0.5451$ mN/μm. Because of the rigid nature of the grasped object, which exhibits a stiffness value higher than that of the gripper arms, the stiffnesses can be considered as the stiffness values of the gripper arms.

The gripping and interaction testing experiments show that force sensor #1 produces different signs of output signals in the two kinds of contact in terms of object gripping and environmental interaction. Hence, whether the contact force is induced by the gripping operation or environmental interaction can be discriminated by monitoring the sign of the sensor #1 output signal. Specifically, if force sensor #1 produces a positive force signal with magnitude larger than a specified threshold value, it can be determined that a gripping force is applied between the gripper right tip and the object. In contrast, if force sensor #1 creates a negative force signal with magnitude over a threshold, it can be concluded that the right tip interacts horizontally or vertically with the environment.

Force sensor #1 provides a fine resolution of 0.314 mN within a smaller force measuring range [0, 138.3 mN]. Meanwhile, force sensor #2 is able to deliver a coarse resolution of 20.512

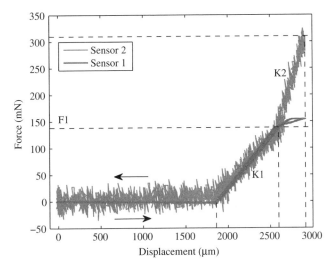

Figure 10.20 Relationship between the gripping force and displacement. The different slopes indicate different stiffnesses K_1 and K_2.

mN within the entire range [0, 815.2 mN]. Hence, in the smaller range [0, 138.3 mN], sensors #1 and #2 provide a redundancy of force measuring. The reason for employing sensor #1 for the force measurement in the smaller range alone is that the piezoresistive strain gauge is very brittle and is vulnerable to break under a large strain. Meanwhile, the metal strain gauge is able to tolerate a larger strain without damage, hence, it is used for the larger-range force sensing and the left-tip position sensing.

The experimental results validate the concept design of the new gripper with dual-sensitivity, dual-range force sensor for both gripping and interaction force sensing. The pick–transport–place experiment demonstrates the efficiency of the gripper in the assembly application. It is notable that the presented gripper system offers fine resolution only in the smaller range. At the limit of the smaller range, the impact between the mover and the stopper may cause vibration of the flexures. The impact can be alleviated by applying a low-speed driving signal as depicted in the experimental results, shown in Fig. 10.13. In the future, a variable-stiffness compliant mechanism without impact effect will be devised to further enhance the gripper performance. In addition, the quarter-bridge circuit is adopted as the signal conditioner of strain-gauge output in this work. The sensitivities of the strain-gauge sensors can be further enhanced by constructing a half- or full-bridge circuit.

10.6 Conclusion

This chapter presents the design, analysis, fabrication, and experimental testing of a large-range compliant gripper with integrated strain-gauge position and force sensors. The gripper provides a gripping range of 4.40 mm with a displacement resolution of 9.724 μm. The dual-sensitivity force sensor produces different resolutions of 0.314 mN and 20.512 mN within two ranges: [0, 138.3 mN] and [0, 815.2 mN], respectively. The gripping range and force sensing ranges can be adjusted to cater for different applications. Experimental results demonstrate that the single force sensor is able to detect the interaction force and gripping force sequentially. The pick–transport–place of a small screw demonstrates the effectiveness of the developed gripper in assembly applications. In addition, different variations of the gripper structure can easily be generated from the proposed one.

References

[1] Lopez-Walle, B., Gauthier, M., and Chaillet, N. (2008) Principle of a submerged freeze gripper for microassembly. *IEEE Trans. Robot.*, **24** (4), 897–902.

[2] Bolopion, A., Xie, H., Haliyo, D.S., and Regnier, S. (2012) Haptic teleoperation for 3-D microassembly of spherical objects. *IEEE/ASME Trans. Mechatron.*, **17** (1), 116–127.

[3] Reddy, A.N., Maheshwari, N., Sahu, D.K., and Ananthasuresh, G.K. (2010) Miniature compliant grippers with vision-based force sensing. *IEEE Trans. Robot.*, **26** (5), 867–877.

[4] Xu, Q. (2012) New flexure parallel-kinematic micropositioning system with large workspace. *IEEE Trans. Robot.*, **28** (2), 478–491.

[5] Du, Z., Shi, R., and Dong, W. (2014) A piezo-actuated high-precision flexible parallel pointing mechanism: Conceptual design, development, and experiments. *IEEE Trans. Robot.*, **30** (1), 131–137.

[6] Krijnen, B. and Brouwer, D.M. (2014) Flexures for large stroke electrostatic actuation in MEMS. *J. Micromech. Microeng.*, **24**, 015 006.

[7] Estevez, P., Bank, J.M., Porta, M., Wei, J., Sarro, P.M., Tichem, M., *et al.* (2012) 6 DOF force and torque sensor for micro-manipulation applications. *Sens. Actuator A: Phys.*, **186**, 86–93.

[8] Ni, Z., Bolopion, A., Agnus, J., Benosman, R., and Regnier, S. (2012) Asynchronous event-based visual shape tracking for stable haptic feedback in microrobotics. *IEEE Trans. Robot.*, **28** (5), 1081–1089.

[9] Kim, K., Liu, X., Zhang, Y., and Sun, Y. (2008) Nanonewton force-controlled manipulation of biological cells using a monolithic MEMS microgripper with two-axis force feedback. *J. Micromech. Microeng.*, **18** (5), 055 013.

[10] Wei, J., Porta, M., Tichem, M., and Sarro, P.M. (2009) A contact position detection and interaction force monitoring sensor for micro-assembly applications, in *Proc. 15th Int. Conf. on Solid State Sensors, Actuators and Microsystems (Transducers 2009)*, Denver, CO, pp. 2385–2388.

[11] Xu, Q. (2013) Design and implementation of a novel compliant rotary micropositioning stage driven by linear voice coil motor. *Rev. Sci. Instrum.*, **84** (5), 055 001.

[12] Xu, Q. (2014) A novel compliant micropositioning stage with dual ranges and resolutions. *Sens. Actuator A: Phys.*, **205**, 6–14.

[13] Leang, K.K., Shan, Y., Song, S., and Kim, K.J. (2012) Integrated sensing for IPMC actuators using strain gauges for underwater applications. *IEEE/ASME Trans. Mechatron.*, **17** (2), 345–355.

11

MEMS Rotary Gripper

Abstract: This chapter presents the design, analysis, fabrication, and testing of a micro-electromechanical systems (MEMS) microgripper with integrated electrostatic actuator and capacitive force sensor. The force sensor is capable of measuring the gripping force and environmental interaction force in two perpendicular axes. The compact gripper structure is designed based on flexure-based compliant rotary bearing and linear guiding mechanisms. To facilitate the parametric design of the gripper, analytical models are established which are verified by conducting finite-element analysis (FEA) simulations. The microgripper is fabricated by the silicon-on-insulator (SOI)-based process. Experimental investigations are carried out to demonstrate the gripping and sensing performances. The effectiveness of the developed gripper device is demonstrated by gripping a human hair through experimental study.

Keywords: Microelectromechanical systems, Robotic grippers, Flexure mechanisms, Compliant mechanisms, Gripping force sensing, Interaction force sensing, Finite-element analysis, Electrostatic actuators, Capacitive force sensors, Silicon-on-insulator.

11.1 Introduction

As a key device in micro-/nanohandling systems, the microgripper plays an important role dedicated to micromanipulation and microassembly tasks [1, 2]. Typically, the gripper is employed to execute a grasp–hold–release operation for tasks such as material characterization, biological sample pick-and-place, and micro-objects assembly, etc. In Chapter 10, a meso-scale compliant gripper with an over 4-mm gripping range was reported for microassembly applications. This chapter presents the miniaturization of the gripper into MEMS scale. Given the gripper structure, which is designed based on a flexure structure using rotary bearing and linear guiding mechanisms, a suitable actuator and sensor should be devised to enable the driving and sensing functions of the gripper, respectively.

Regarding the driving approach, different types of actuators have been employed to actuate the microgripper. For example, piezoelectric actuation [3, 4], pneumatic actuation [5, 6], electromagnetic actuation [7, 8], electrothermal actuation [9–11], and electrostatic actuation [12–15] have been proposed. Generally, different actuators endow the gripper with different performances. In particular, piezoelectric actuators introduce hysteretic nonlinearity to

Design and Implementation of Large-Range Compliant Micropositioning Systems, First Edition. Qingsong Xu.
© 2016 John Wiley & Sons Singapore Pte Ltd. Published 2016 by John Wiley & Sons Singapore Pte Ltd.

the gripper displacement, which demands a sophisticated control design to deliver a precise gripping motion. The work of electromagnetic actuators relies on an external magnetic field, which usually possesses a bulky dimension. Electrothermal actuators are able to provide a large displacement under a relatively low voltage. However, the induced high temperature during operation restricts their applications in manipulating temperature-sensitive materials. In contrast, electrostatic actuators are capable of offering sufficient displacements for gripping samples in both ambient and liquid environments [16]. Hence, the electrostatic actuator is employed to drive the gripper in this work.

To grasp micro-objects delicately without damaging the gripper and the grasped objects, it is crucial to detect and regulate the gripping force between the gripper tips and the objects. Therefore, the force sensing capability is required to cater for this requirement. In the literature, the electrothermal [17], piezoresistive [13], and capacitive [14, 16] types of force sensors have been realized in microgrippers. Considering the relatively high resolution, the capacitive sensors are adopted in this work to enable the force sensing function of the microgripper.

Usually, the force sensor is employed to detect the gripping force between the gripper tips and the grasped object. Moreover, to guarantee a reliable manipulation, it is also necessary to detect the interaction force exerted by the environment on the gripper tip [18]. The interaction force sensing is required to determine the exact moment when the contact is established between the gripper tip and the environment. Such a function is important to prevent the gripper device from exerting excessive interaction. However, relatively few works have been conducted to enable the interaction force sensing capability of a microgripper in the literature. In addition, the existing grippers require two separate force sensors for the grasping and interaction force sensing [9, 19], which complicates the gripper design process and increases the hardware costs.

To overcome the aforementioned issues, a new MEMS microgripper with integrated electrostatic actuator and capacitive force sensor is presented in this chapter. In particular, a single force sensor is designed to measure the grasping and interaction forces in two perpendicular directions.

11.2   MEMS Gripper Design

Figure 11.1 shows a schematic diagram of the designed gripper with integrated actuator and sensor. The compliant gripper is composed of a left actuation tip and a right force sensing tip. Each gripper tip is connected to a multi-stage compound radial flexure (MCRF)-based rotary arm bearing. The actuation arm bearing is connected to the lateral comb-drive electrostatic actuator, which is supported by four actuation flexures. The sensing arm bearing is linked to the transverse comb-drive capacitive force sensor, which is supported by four sensing flexures. The actuation and sensing flexures act as linear guides for the actuator and sensor, respectively. It is notable that the actuator and sensor segments are divided and contact electrodes are formed in this device layer for the SOI microfabrication, as shown in Fig. 11.4 later.

Referring to Fig. 11.1, it is observed that the role of the actuation arm bearing is to convert the actuator's displacement in the y-axis direction into the gripping motion of the left tip along the x-axis direction. Similarly, the role of the sensing arm bearing is to transmit the x-axis motion of the right tip into the y-axis translation of the force sensor plates. In this way, the gripping force F_g is measured by the force sensor. It is notable that the two rotary bearings belong to MCRFs with $N = 1$, as presented in Chapter 8.

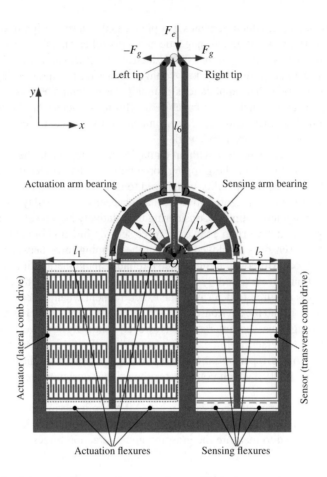

Figure 11.1   Schematic diagram of a microgripper with the integrated actuator and sensor.

In addition, the right tip is slightly longer than the left one. Therefore, when the gripper interacts with the environment in the y-axis direction, contact is established with the right tip first. The interaction force F_e excites a y-axis motion of the right tip, which is also transferred to the force sensor flexures. Thus, both the gripping and interaction forces can be detected by the single force sensor.

It is notable that the gripper tip does not deliver a pure translational motion, as it is guided by a rotary bearing flexure. Actually, to achieve a gripping range up to several hundred micrometers, the slight rotation of the gripper tips is negligible.

11.2.1   Actuator Design

11.2.1.1   Driving Issues

The electrostatic actuators have two different types, i.e., lateral comb drive and transverse comb drive. Generally, the former is able to produce a larger displacement than the latter. Hence, the lateral comb drive is selected to actuate the gripper in this work.

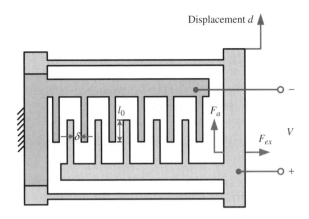

Figure 11.2   Illustration of the lateral comb-drive actuator.

A schematic presentation of the lateral comb-drive actuator is shown in Fig. 11.2. The comb tooth electrodes are considered as parallel plates. The driving force of the actuator is determined by

$$F_a = \frac{\epsilon b V^2 N_c}{\delta} \tag{11.1}$$

where $\epsilon = 8.85 \times 10^{-12}$ C^2/(N · m^2) is the permittivity of air, δ is the gap width between two adjacent plates, b is the out-of-plane thickness of the gripper plate, V is the driving voltage, and N_c is the number of comb tooth pairs. For example, Fig. 11.2 shows a comb drive with five pairs of comb teeth (i.e., $N_c = 5$).

Considering the parallel connections of the four actuation flexures as denoted in Fig. 11.1, the equivalent stiffness can be calculated by

$$K_1 = \frac{4Ebh_1^3}{l_1^3} \tag{11.2}$$

where E is the Young's modulus of silicon, and h_1 and l_1 are the in-plane width and length of the leaf flexures, respectively.

The torsional spring constant of the actuation arm bearing is calculated in view of the stiffness equation (8.13) of the MCRF with $N = 1$:

$$K_{t2} = \frac{Ebh_2^3 r_1 R_1}{l_2^3} \tag{11.3}$$

where h_2 and l_2 are the in-plane width and length of the radial flexures, r_1 is the inner radius, and $R_1 = r_1 + l_2$ as denoted in Fig. 11.1.

Considering that a tangential driving force F_A produces an angle θ of the rotary bearing, the torsional stiffness can also be expressed as follows:

$$K_{t2} = \frac{F_A l_5}{\theta} \tag{11.4}$$

where l_5 is the driving distance as shown in Fig. 11.1.

In addition, the rotary angle θ of the bearing induces a translation d of the driving point A, which can be written approximately as

$$d = l_5\theta. \tag{11.5}$$

Hence, in view of Eqs. (11.3), (11.4), and (11.5), the equivalent linear stiffness of the rotary bearing at the driving point A can be derived as

$$K_2 = \frac{F_A}{d} = \frac{K_{f2}}{l_5^2} = \frac{Ebh_2^3 r_1 R_1}{l_2^3 l_5^2}. \tag{11.6}$$

Considering that the spring constants K_1 and K_2 are connected in parallel because both are fixed at the base, the equivalent actuation stiffness for the left arm of the gripper can be computed as

$$K_a = K_1 + K_2 = Eb\left(\frac{4h_1^3}{l_1^3} + \frac{h_2^3 r_1 R_1}{l_2^3 l_5^2}\right) \tag{11.7}$$

which is derived in view of Eqs. (11.2) and (11.6).

Taking into account Eqs. (11.1) and (11.7), the driving displacement can be determined as

$$d = \frac{F_a}{K_a} = \frac{\epsilon V^2 N_c}{E\delta}\left(\frac{4h_1^3}{l_1^3} + \frac{h_2^3 r_1 R_1}{l_2^3 l_5^2}\right)^{-1}. \tag{11.8}$$

The driving displacement d of the actuator will be transmitted as the displacement of the output point C as shown in Fig. 11.1. The produced gripping range of the left tip can be calculated as

$$x_g = A_x d \tag{11.9}$$

where the displacement amplification ratio A is given by

$$A_x = \frac{l_5 + l_6}{l_5} \tag{11.10}$$

with l_6 denoting the length of the gripper arm.

It is observed from Eq. (11.8) that the larger the length l_1 of the actuation flexure and the length l_2 of the actuation arm bearing, the larger the driving displacement d, and hence the larger the gripping range x_g of the left tip as revealed by Eq. (11.9). Thus, the flexure length can be determined according to the gripping range requirement. In practice, the maximum length is constricted by the fabrication process requirement.

At the same time, the driving force F_a of the actuator is transmitted as the gripping force F_g of the left tip as follows:

$$F_g = \frac{F_a}{A_x}. \tag{11.11}$$

In view of Eqs. (11.9) and (11.11), the stiffness of the left arm seen at the left tip is determined by

$$K_g = \frac{F_g}{x_g} = \frac{F_a}{A_x^2 d} = \frac{K_a}{A_x^2} = \frac{Eb}{A_x^2}\left(\frac{4h_1^3}{l_1^3} + \frac{h_2^3 r_1 R_1}{l_2^3 l_5^2}\right). \tag{11.12}$$

11.2.1.2 Side Instability Issues

In addition to producing the lateral driving force F_a, the comb-drive actuator also creates an electrostatic force F_{ex} in the transverse direction, as shown in Fig. 11.2. Due to the effect of F_{ex}, the movable teeth are pulled toward the fixed teeth of the actuator during the drive. This phenomenon is called the side instability of comb drives.

As shown in the literature [20, 21], the electrostatic force F_{ex} can be calculated as

$$F_{ex} = \frac{N_c \epsilon b (d + l_0) V^2}{2} \left[\frac{1}{(\delta - x)^2} - \frac{1}{(\delta + x)^2} \right] \tag{11.13}$$

where l_0 is the initial overlap length of the comb-drive teeth and x is the translation of the movable teeth in the transverse direction along the x-axis.

When the movable teeth are located at the center of the gap, the magnitude of the equivalent negative stiffness is given by

$$K_n = \left. \frac{\partial F_{ex}}{\partial x} \right|_{x=0} = \frac{2 N_c \epsilon b (d + l_0) V^2}{\delta^3}. \tag{11.14}$$

It is observed from Eq. (11.14) that the larger the input displacement d, the larger the stiffness K_n. To avoid the side instability, the transverse stiffness K_x of the structure should be greater than K_n. Hence, the side instability occurs when $K_n = K_x$; that is,

$$\frac{2 N_c \epsilon b (d + l_0) V^2}{\delta^3} = \frac{4 E b h_1}{l_1} = K_x. \tag{11.15}$$

Substituting Eq. (11.8) into Eq. (11.15), a necessary algebra operation allows the calculation of the allowable maximum driving voltage of the electrostatic actuator without side sticking:

$$V_{max}^2 = \frac{\delta^2 K_a}{2 N_c \epsilon b} \left(\sqrt{\frac{2 K_x}{K_a} + \frac{l_0^2}{\delta^2}} - \frac{l_0}{\delta} \right). \tag{11.16}$$

11.2.2 Sensor Design

11.2.2.1 Gripping Force Sensor

Referring to Fig. 11.1, when the object is grasped by the two tips, the gripping force exerted on the right tip will be transmitted to the movable plates of the capacitive sensor. The capacitive sensor is constructed by transverse comb drives, and the sensing principle is illustrated in Fig. 11.3.

Here, the output signal V_{out} of the capacitive sensor is produced by a capacitive-to-voltage converter chip (model: MS3110, from MicroSensors, Inc.):

$$V_{out} \propto G_{ain} \cdot \frac{C_1 - C_2}{C_F} \tag{11.17}$$

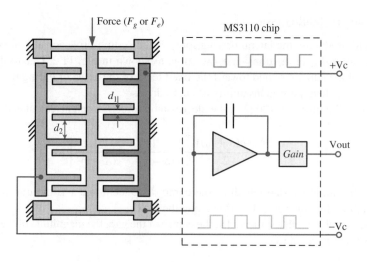

Figure 11.3   Capacitive sensing scheme with readout circuit.

with

$$C_1 = \frac{\epsilon A}{d_1 - y} + \frac{\epsilon A}{d_2 + y} \tag{11.18}$$

$$C_2 = \frac{\epsilon A}{d_1 + y} + \frac{\epsilon A}{d_2 - y} \tag{11.19}$$

where G_{ain} and C_F are adjustable parameters and A is the area of the capacitor plates. d_1 and d_2 are the smaller and larger gaps of the capacitor plates, respectively, as shown in Fig. 11.3. It has been shown [9] that the output voltage V_{out} is in proportion to the deflection y.

Similar to the analysis of the left arm, the equivalent stiffness of the right arm at the point D can be calculated as

$$K_s = \frac{F_s}{d} = K_3 + K_4$$

$$= \frac{4Ebh_3^3}{l_3^3} + \frac{Ebh_4^3 r_2 R_2}{l_4^3 l_5^2}$$

$$= Eb\left(\frac{4h_3^3}{l_3^3} + \frac{h_4^3 r_2 R_2}{l_4^3 l_5^2} \right) \tag{11.20}$$

where r_2 is the inner radius parameter and $R_2 = r_2 + l_4$ as shown in Fig. 11.1.

Regarding the right arm of the gripper, the displacement x_f at the right tip is deamplified as the displacement x_s at point B. That is,

$$x_s = \frac{x_f}{A_x} \tag{11.21}$$

where the amplification ratio A_x is calculated from Eq. (11.10).

In addition, the gripping force F_g of the right tip is transmitted as the sensing force F_s of the force sensor at the point B:

$$F_s = A_x F_g. \tag{11.22}$$

In consideration of Eqs. (11.21) and (11.22), the equivalent stiffness observed at the right tip is given by

$$K_f = \frac{F_g}{x_f} = \frac{F_s}{x_s A_x^2} = \frac{K_s}{A_x^2}. \tag{11.23}$$

In practice, when an object is grasped by driving the left tip, the gripping force causes a deflection of the right tip in the x-axis direction. To induce a smaller deflection of the sensing tip, it is desirable to make the right tip stiffer than the left one. That is,

$$K_f > K_g. \tag{11.24}$$

To measure the gripping force F_g using the force sensor, the caused deflection of the right tip is determined by

$$x_f = \frac{F_g}{K_f} \tag{11.25}$$

which will cause the movable plates of the force sensor to translate a distance x_s as calculated from Eq. (11.21).

Given the allowable maximum one-sided translation x_s^{max} of the force sensor plates, the measured maximum gripping force can be determined as follows:

$$F_g^{max} = K_e A_x x_s^{max}. \tag{11.26}$$

11.2.2.2 Environmental Interaction Force Sensor

During a typical micromanipulation task, before the gripping operation of micro-objects, the gripper is usually transported by a micropositioning stage to be adjacent to the object or environment. For instance, by determining the position where an initial contact is made with the supporting base, the position of the gripper tip with respect to the objects to be grasped can be calibrated precisely.

When an environmental interaction force F_e is exerted on the right tip as shown in Fig. 11.1, this force will be transmitted to the end point B of the force sensor. The force causes a translation x_e of the movable plates of the sensor. Considering that F_e is transmitted to the force sensor directly, the stiffness K_e of the right arm of the gripper in the y-axis direction is approximated by K_s, as shown in Eq. (11.20).

Given the maximum allowable translation x_s^{max} of the force sensor, the maximum interaction force that can be measured is determined by

$$F_e^{max} = K_e x_s^{max}. \tag{11.27}$$

11.3 Performance Evaluation with FEA Simulation

As a case study, the structural parameters of the gripper are designed to provide a gripping range of 80 μm at a driving voltage of 85 V. The main parameters of the gripper are shown in Table 11.1. The driving voltage is far less than the allowable maximum driving voltage of 322 V as predicted by Eq. (11.16). In addition, the allowable maximum one-sided translation for the movable plates of the force sensor is designed as 5 μm. Hence, the force sensor is expected

Table 11.1 Main parameters of a MEMS microgripper

Parameter	Symbol	Value	Unit
Teeth overlap length	l_0	40	μm
Actuation flexure length	l_1	1000	μm
Actuation flexure width	h_1	7	μm
Arm bearing length	l_2, l_4	500	μm
Arm bearing width	h_2, h_4	7	μm
Sensing flexure length	l_3	600	μm
Sensing flexure width	h_3	10	μm
Driving distance	l_5	970	μm
Arm length	l_6	2050	μm
Inner radius of arm bearing	r_1, r_2	310	μm
Gap 1 of capacitor plates	d_1	5	μm
Gap 2 of capacitor plates	d_2	20	μm
Gap of teeth	δ	5	μm
Device layer thickness	b	50	μm
Number of comb tooth pairs	N_c	594	–

to measure a maximum gripping force of 200 μN. A CAD model of the designed gripper is depicted in Fig. 11.4. The performance of the gripper structure is evaluated by conducting FEA simulation studies as follows.

11.3.1 Statics Analysis

To evaluate the statics performance of the gripper, a static structural FEA is carried out with the ANSYS software package. To speed up the simulation, the comb-drive teeth and capacitor plates are removed, which does not influence the static structural analysis result. Under a driving force $F_a = 100$ μN, the simulation result of the gripper deformation is shown in Fig. 11.5. By extracting the input and output displacements of the gripper, the amplification ratio A_x and stiffness K_g of the left arm are derived. In addition, by applying a gripping force F_g and an interaction force F_e on the tip of the right arm along the x-axis and y-axis directions, respectively, the induced displacements are obtained. Then, the stiffness values K_f and K_e of the right arm are calculated.

For the purpose of comparison, the results obtained by the analytical models in Eqs. (11.10), (11.12), (11.20), and (11.23), and FEA simulations, are presented in Table 11.2. Taking the FEA results as the benchmark, it is observed that the errors of the analytical models are within 25%. The model errors mainly come from the assumption employed in the analytical models, where only the bending deformations of leaf flexures are considered. The model accuracy can be improved by taking into account all types of flexure deformations.

FEA simulation results show that the stiffness K_f of the sensing arm is over seven times larger than the stiffness K_g of the actuation arm. This means that the right tip will experience a deflection which is less than one seventh that of the left tip during the gripping operation. In addition, the interaction stiffness K_e is larger than the gripping stiffness K_f, which means that the sensing arm can accommodate a larger interaction force than the gripping force, as can be deduced from Eq. (11.27).

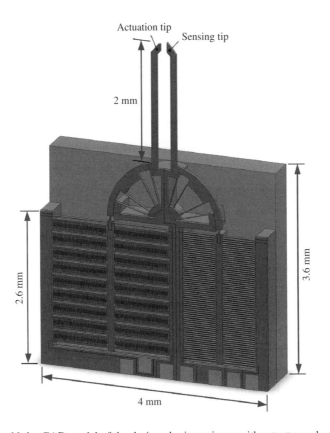

Figure 11.4 CAD model of the designed microgripper with actuator and sensor.

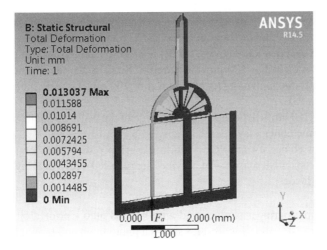

Figure 11.5 FEA simulation result of the gripper structure with an actuation force F_a applied.

Table 11.2 Analytical model and FEA simulation results of the gripper structure performance

Performance	Model	FEA	Error (%)
Amplification ratio A_x	3.11	3.13	0.6
Stiffness of left arm K_g ($\mu N/\mu m$)	1.41	1.64	14.0
Stiffness of right arm K_f ($\mu N/\mu m$)	12.91	11.71	10.2
Stiffness of right arm K_e ($\mu N/\mu m$)	125.13	167.07	25.1

11.3.2 Dynamics Analysis

The dynamics performance of the gripper structure is examined by conducting modal analysis simulation. Similarly, to speed up the simulation, the comb-drive teeth and capacitor plates are removed, while the masses of their moving parts are added to the linear shuttles of the left and right arms, respectively.

The simulation results of the first two resonant modes are shown in Fig. 11.6. It is observed that the first and second resonant frequencies are 1430.4 Hz and 3877.9 Hz, which are attributed to the actuation and sensing arms, respectively.

11.4 Gripper Fabrication

The designed gripper is fabricated using a SOI wafer with a device layer of 50 μm. Referring to Fig. 11.7, the microfabrication procedure is presented as follows.

(a) A 4-inch, P-type, < 100 >-orientation SOI wafer with a device layer of 50 μm, handling layer of 400 μm, and a silicon oxide (SiO2) interlayer of 2 μm is used as the starting substrate in the fabrication.
(b) Thermal oxidation is used to deposit a SiO2 layer of 1.5 μm on the bottom side of the wafer. A layer is lithographically patterned, and reactive ion etching (RIE) is used to remove the remaining oxide layer on the bottom side.
(c) Backside deep reactive ion etching (DRIE) is used to etch the handling layer. When the handling layer is etched by 200 μm, the patterned SiO2 in step (b) is removed. Then, DRIE is applied to etch the remaining handling layer of 200 μm. The SiO2 interlayer of 2 μm is etched through RIE.
(d) An Al layer of 250 nm is deposited on the top side of the wafer by sputtering, and patterned to form an ohmic contact layer.
(e) The top side is etched using DRIE to form the gripper structure along with comb-drive actuators and capacitive sensors. The devices are released onto a dummy wafer which acts as the support layer.

The fabricated microgripper is glued and wire-bonded onto a printed circuit board (PCB) for experimental investigation. Figure 11.8 shows a photograph and scanning electron microscope (SEM) images of different components of the fabricated prototype gripper. The planar size of the microgripper is 4.0 mm × 5.6 mm. The initial clearance between the two tips of the gripper is 120 μm. Additionally, the right tip is designed to be slightly longer than the left one

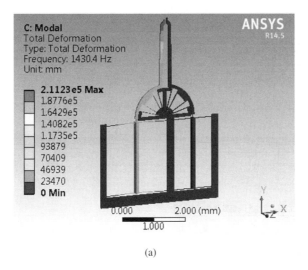

(a)

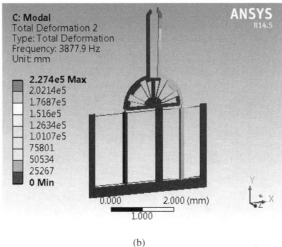

(b)

Figure 11.6 FEA simulation results of the modal analysis for the gripper structure: (a) mode 1; (b) mode 2.

to facilitate the interaction force sensing and calibration process of the force sensor. In the subsequent section, the performance of the fabricated microgripper is tested through experimental studies.

11.5 Experimental Results and Discussion

11.5.1 Gripping Range Testing Results

The gripping range of the microgripper is tested by experimental study first. The overall experimental setup is shown in Fig. 11.9. The gripper is mounted on an XYZ micromanipulator

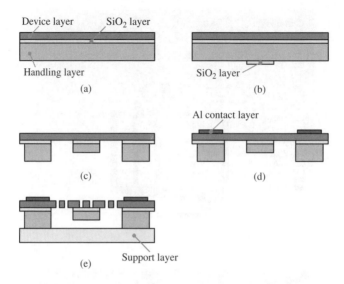

Figure 11.7   Fabrication processes of the MEMS gripper.

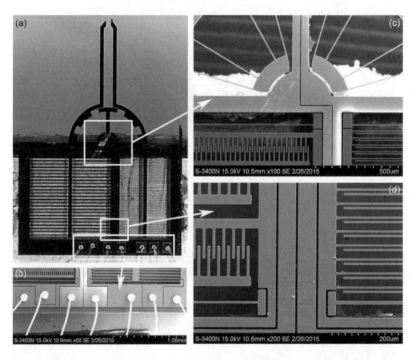

Figure 11.8   (a) Photograph and (b)–(d) scanning electron microscopy images of the fabricated micro-gripper prototype.

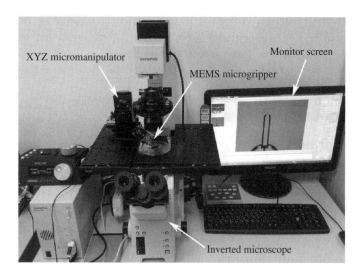

Figure 11.9 Experimental setup for performance testing of the microgripper under an inverted microscope.

(model: MP-285, from Sutter Instrument Inc.) and put under an inverted microscope (model: IX81, from Olympus Corp.).

Driving the left arm of the gripper by applying different voltages, the images of the gripper tips are acquired and then processed to extract the displacement of the left tip. The results are

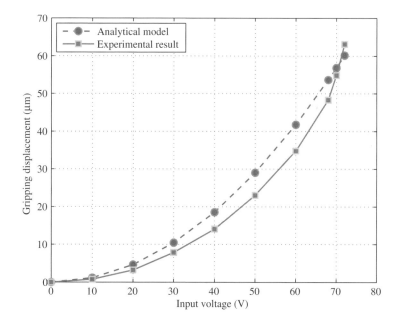

Figure 11.10 Gripping displacement of the microgripper versus the driving voltage.

shown in Fig. 11.10. It is observed that the gripper provides a gripping range of 63 μm with a driving voltage of 72 V. The experimental results agree with the analytical model prediction results, with a certain deviation. The discrepancy may arise from the fabrication tolerance of the gripper parameters, due to the over-etching effect and the modeling error of the structure stiffness.

It is found that a driving voltage higher than 72 V leads to the side-sticking phenomenon of the comb-drive actuator, which causes a side instability of the electrostatic actuator. Compared with the maximum allowable voltage of 322 V predicted by Eq. (11.16), the actual value is much smaller. The theoretical reason is examined by inspecting the FEA simulation results of the gripper structure when a driving force is applied.

Referring to Fig. 11.5, it is observed that when the actuator produces a driving displacement of 63 μm, the transverse displacement of the actuator along the x-axis is about $x_m = 50$nm. Such a non-negligible transverse translation causes the early side sticking of the movable comb teeth with the fixed ones. This transverse displacement is mainly induced by the x-axis translation of the point A of the actuation bearing flexure. It is notable that the intrinsic displacement x_m is amplified by the lateral electrostatic forces and produces an extra displacement x_e. The total lateral displacement in the x-axis direction is $x_m + x_e$, as reported in the literature [21]. In addition, the actual transverse displacement x_m may be larger than the simulation result of 50 nm due to fabrication imperfections, which will also contribute to the early side sticking of the comb drives.

In order to mitigate the side-sticking effect, one possible way is to increase the transverse stiffness of the tethering beams (e.g., change the current fixed–guided tethering beams of the comb-drive actuator to folded beams) so as to reduce the transverse displacement of the movable comb-drive teeth.

Other possible reasons that cause the above discrepancy may arise from the stiffness prediction equations for K_x and K_a and the fabrication imperfection of the structural parameters of the gripper.

11.5.2   Gripping Force Testing Results

In order to test the gripping force of the gripper, the force sensor is calibrated using a commercially available microforce sensing probe (model: FT-S100, from FemtoTools AG). The experimental setup is shown in Fig. 11.11. The sensing probe provides a resolution of 0.05 μN within a measuring range of ±100 μN. To apply a force to the right tip of the microgripper, the microforce sensing probe is fixed on an XY nanopositioner (model: P-734.2CL, from Physik Instrumente Co., Ltd.), which enables an in-plane translation of the probe with a resolution of 0.3 nm in the travel range of 100 μm × 100 μm. In addition, the microgripper is fixed on an XYZ micromanipulator (model: MP-285, from Sutter Instrument Inc.), which allows the adjustment of the gripper position relative to the microforce sensing probe in three axes.

To calibrate the force sensor of the microgripper, an experimental study is conducted as follows. First, the microgripper is positioned adjacent to the microforce probe. Next, the probe is driven to translate a forward distance of 3 μm with a step size of 50 nm to make contact with the gripper's right tip. Then, the probe is moved backward. The calibration result is shown in Fig. 11.12. The steps in the ascending and descending curves are caused by the consecutive-step motion of the nanopositioner with a 50-nm step size. It is observed that the

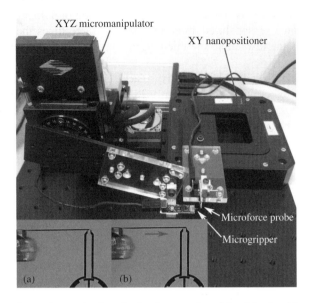

Figure 11.11 Experimental setup for the gripping force testing of the right tip with a microforce sensing probe. The close-up views are (a) before and (b) after the contact.

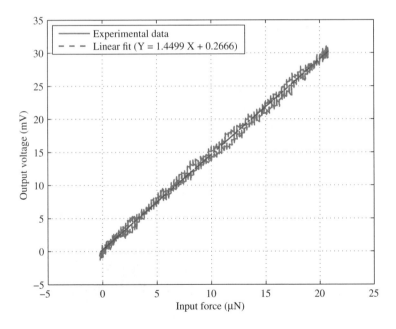

Figure 11.12 Output voltage of the gripping force sensor versus the input force.

force sensor of the gripper exhibits slight hysteresis error. Based on the experimental data, a linear regression function is fitted, which exhibits a slope of 1.45 mV/μN, which is the sensitivity of the force sensor. In addition, by adopting three times the standard deviation (3σ) of the noise as the resolution, the microgripper provides a force sensing resolution of 0.61 μN.

Taking into account the induced force and displacement by the nanopositioner, the stiffness of the sensing tip of the gripper is obtained as 9.28 μN/μm. Compared with the FEA result of 11.71 μN/μm, the experimental result is 20% lower. The discrepancy is mainly caused by the difference between the designed and actual parameters of the flexure beams.

It is notable that the displacement amplification ratio between the right tip and the point B shown in Fig. 11.1 is $A = 3.1$. In order to excite the maximum displacement of ±5 μm for the movable plates of the capacitive force sensor, the right tip of the gripper should translate a distance of ±5 μm × 3.1 = ±15.55 μm. Hence, the measuring range of the gripping force sensor can be determined as ±15.55 μm × 9.28 μN/μm = ±144.4 μN.

11.5.3 Interaction Force Testing Results

The interaction force sensing capability of the microgripper is tested by using the experimental setup shown in Fig. 11.13. Similar to the setup depicted in Fig. 11.11, the microgripper is mounted on an XYZ micromanipulator. The microforce sensing probe is carried by an XY

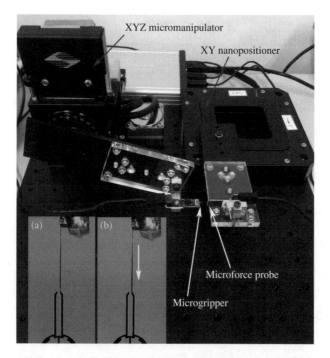

Figure 11.13 Experimental setup for the interaction force testing of the right tip with a microforce sensing probe. The close-up views are (a) before and (b) after the contact.

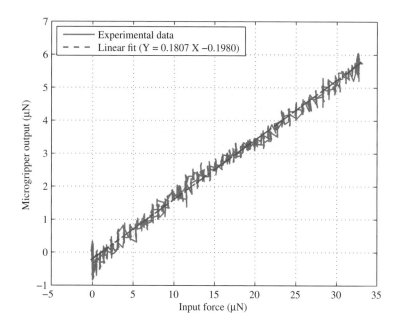

Figure 11.14 Interaction force sensor output of the microgripper versus the input force.

nanopositioner, which translates with a smaller step size of 5 nm to exert the interaction force F_e on the right tip of the gripper along the y-axis.

The force signals produced by the commercial force probe and the microgripper are recorded as shown in Fig. 11.14. It is observed that the force sensor of the gripper produces a smaller magnitude of force than the microforce probe. A linear regression function is fitted from the experimental data, which exhibits a slope of 0.1807 µN/µN. This indicates that the force readings of the microgripper are only about 1/5.5 of the "true" values produced by the force probe. The reason lies in the fact that the force sensor of the microgripper is calibrated for the gripping force sensing. For the interaction force sensing, the force sensor of the gripper exhibits a 5.5 times lower sensitivity.

The different sensitivities of the force sensing in the x-axis gripping and y-axis interaction directions are mainly caused by the different translational displacements of the sensing point B, which are excited by the same force magnitude applied to the gripper tip in the gripping and interaction directions. In other words, with the same force applied in the gripping and interaction directions (i.e., $F_g = F_e$) the interaction force F_e induces a smaller displacement of the sensor plates due to a higher stiffness. Subsequently, it causes a smaller value of the force sensor reading and a lower sensitivity of the force sensing in the interaction direction.

11.5.4 Demonstration of Micro-object Gripping

To demonstrate the performance of the developed microgripper, the gripper is employed to grasp a human hair under an inverted microscope (model: IX81, from Olympus Corp.). The experimental setup is shown in Fig. 11.15. The gripper is fixed on an XYZ micromanipulator (model: MP-285, from Sutter Instrument Inc.). The hair has a diameter of about 76 µm and is

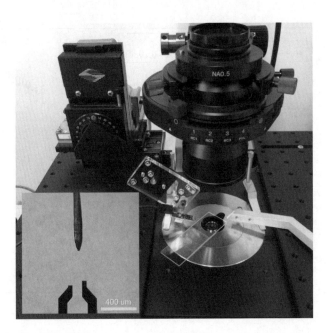

Figure 11.15 Experimental setup for human hair gripping of the microgripper under an inverted microscope.

fixed horizontally. Before gripping testing, the position of the microgripper relative to the hair is adjusted using the XYZ micromanipulator. Then, the gripper is driven to grasp and release the hair.

Starting from the initial position (I) in the gripping process, the left tip of the gripper is first driven to translate 44 μm. Then, it contacts the hair (position II). Afterwards, the gripping force is gradually increased to 28 μN. The gripping holds for 10 s, and then the hair is released by reducing the driving voltage. Subsequently, the gripping force is reduced gradually. At the end of the release operation (position III), the force sensor exhibits a negative gripping force up to –3.5 μN. The driving voltage and the gripping force are recorded, and the relationship between the gripping force and the input voltage is shown in Fig. 11.16. The negative force is caused by the adhesive force between the gripper tip and the hair.

In addition, Fig. 11.16 exhibits a deviation between the ascending and descending curves. This discrepancy is caused by a slight drift of the gripping force during the loading and unloading processes. The force drift may arise from different sources. One possible reason lies in the permanent deformation of the grasped hair. Because a constant voltage is commanded to apply the holding force during the 10-s gripping operation, the gripping force will change slightly if a permanent deformation of the hair appears. In addition, considering that the experiments are conducted in room conditions, the drift may also be caused by the influence of external disturbances on the capacitive sensor due to the temperature and flow variations.

11.5.5 Further Discussion

The foregoing experimental results demonstrate the gripping and force sensing performances of the developed microgripper. The same force sensor is designed to function as a gripping

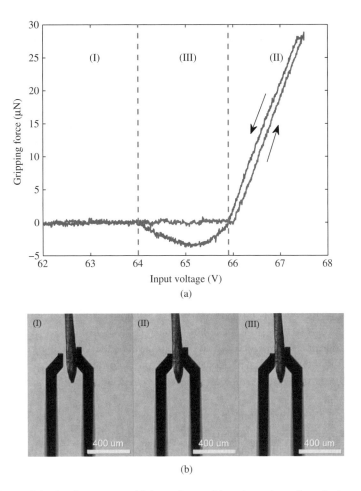

Figure 11.16 (a) Gripping force versus driving voltage of the microgripper for gripping a human hair; (b) images for three positions.

force sensor and an interaction force sensor in an alternate manner. However, when acting as an interaction force sensor, the sensor offers a worse sensitivity – 5.5 times lower than that of the gripping force sensing. Even so, the experimental results validate the proposed conceptual design of the microgripper with a single force sensor, which provides an alternative, cost-effective method compared with the existing microgrippers adopting two individual force sensors [9, 19].

The reported gripper can be employed in various micromanipulation tasks. For example, in microassembly processes, before the gripping operation, the force sensor of the gripper can be used as an interaction force sensor to detect contact with the environment. Afterwards, the same sensor can be employed as a gripping force sensor for the grasp force sensing. It is notable that the force sensor is designed to detect the two axial forces separately. If the gripping and interaction forces are applied simultaneously, the force sensor cannot discriminate between the force components. Such a function can be realized by different approaches. For instance, a force sensor can be added to the left arm of the gripper. When the two kinds of forces are exerted at the same time, the difference between the right- and left-arm sensors will give the interaction force.

As a case study, the initial clearance between the two tips of the normally open gripper is designed as 120 μm in this work. Considering the gripping range of the left tip, the gripper can be employed to grasp micro-objects sized between 57 μm and 120 μm. Actually, the clearance between the two tips of the gripper can be designed as any size to facilitate the gripping of target micro-objects. It is notable that the gripper can be employed to handle either soft or rigid objects. When gripping a soft tissue or biological cell which is deformable, the gripping force can be detected by the force sensor as long as the force magnitude is greater than the force resolution (0.61 μN) of the sensor.

During fabrication, the device layer as shown in Fig. 11.1 is partitioned and etched to form the actuator and sensor segments, as illustrated in Fig. 11.4. The two gripper tips are mechanically connected to and electrically isolated from the actuator and sensor parts at points A and B, respectively. Such a mechanical connection and electrical isolation design allows the operation of the gripper tips in a liquid environment, as shown in the literature [16, 22].

The performances of typical MEMS microgrippers, which are actuated by electrostatic actuators, are tabulated in Table 11.3. It is observed that compared with the microgrippers without force sensing, a force sensing microgripper usually exhibits a much larger planar size. For a batch of microgrippers fabricated out of a silicon wafer, the larger the gripper dimension, the smaller the number of fabricated devices. Therefore, in order to increase the yield rate and reduce the unit cost of the microgripper, it is necessary to design a MEMS gripper with smaller planar dimension.

In the future, an optimization design of the gripper structure will be conducted to reduce the difference between the two axial sensitivities of the force sensor. In addition, a closed-loop force control algorithm will be implemented to adjust the gripping force precisely.

11.6   Conclusion

A MEMS microgripper with integrated electrostatic actuator and capacitive force sensor has been designed, simulated, fabricated, and experimentally tested in this chapter. Benefiting from the structure design, the single force sensor enables the gripping and interaction force sensing in two perpendicular directions. The gripper delivers a gripping range of 63 μm at a relatively low driving voltage of 72 V. The force sensor provides a resolution of 0.61 μN within a sensing range of ±144 μN for the gripping force sensing. Acting as the interaction force sensor, the sensor offers a 5.5 times lower sensitivity than that of the gripping force sensing. The performance of the fabricated microgripper is demonstrated by gripping a human hair of 76 μm diameter. Results show that the force sensing gripper is capable of detecting the gripping and adhesive forces during the gripping/release operation. In the future, the gripper structure will be further improved to enhance the side pull-in stability, and an architectural optimization will be carried out to enhance the sensitivity of the force sensing in the interaction direction.

Table 11.3 Performances of typical microgrippers driven by electrostatic actuators

Planar size (mm^2)	Input voltage (V)	Displacement (μm)	Force sensing principle	Force range (μN)	Sensitivity (mV/μN)	Resolution (μN)	Reference
1.25 × 3.3 = 4.125	80	20	None	None	None	None	[12]
2.9 × 2.7 = 7.830	100	94	None	None	None	None	[15]
6.2 × 3.5 = 21.70	80	50	Piezoresistive	±1200	0.072	3	[13]
4.0 × 5.6 = 22.40	72	63	Capacitive	±144	1.46	0.605	This work
5.0 × 6.5 = 32.50	50	17	Capacitive	200	–	–	[14]
7.7 × 5.6 = 43.12	150	100	Capacitive	±360, ±2800	4.41, 0.55	0.07, 0.52	[16]
8.5 × 7.1 = 60.35	80	90	Electrothermal	17	588	–	[17]

References

[1] Chan, B.D., Mateen, F., Chang, C.L., Icoz, K., and Savran, C.A. (2012) A compact manually actuated micromanipulator. *J. Microelectromech. Syst.*, **21** (1), 7–9.

[2] Tamadazte, B., Paindavoine, M., Agnus, J., Petrini, V., and Piat, N.L.F. (2012) Four d.o.f. piezoelectric microgripper equipped with a smart CMOS camera. *J. Microelectromech. Syst.*, **21** (2), 256–259.

[3] Park, J. and Moon, W. (2003) A hybrid-type micro-gripper with an integrated force sensor. *Microsyst. Technol.*, **9** (8), 511–519.

[4] Xu, Q. (2013) Adaptive discrete-time sliding mode impedance control of a piezoelectric microgripper. *IEEE Trans. Robot.*, **29** (3), 663–673.

[5] Ok, J., Lu, Y.W., and Kim, C.J. (2006) Pneumatically driven microcage for microbe manipulation in a biological liquid environment. *J. Microelectromech. Syst.*, **15** (6), 1499–1505.

[6] Alogla, A., Scanlan, P., Shu, W., and Reuben, R.L. (2012) A scalable syringe-actuated microgripper for biological manipulation. *Procedia Eng.*, **47**, 882–885.

[7] Ger, T.R., Huang, H.T., Chen, W.Y., and Lai, M.F. (2013) Magnetically-controllable zigzag structures as cell microgripper. *Lab Chip*, **13**, 2364.

[8] Kuo, J.C., Tung, S.W., and Yang, Y.J. (2013) A hydrogel-based intravascular microgripper manipulated using magnetic fields, in *Prof. 17th Int. Conf. on Solid-State Sensors, Actuators and Microsystems*, pp. 1683–1686.

[9] Kim, K., Liu, X., Zhang, Y., and Sun, Y. (2008) Nanonewton force-controlled manipulation of biological cells using a monolithic MEMS microgripper with two-axis force feedback. *J. Micromech. Microeng.*, **18** (5), 055 013.

[10] Duc, T.C., Lau, G.K., Creemer, J.F., and Sarro, P.M. (2008) Electrothermal microgripper with large jaw displacement and integrated force sensors. *J. Microelectromech. Syst.*, **17** (6), 1546–1555.

[11] Mackay, R.E., Le, H.R., and Keatch, R.P. (2011) Design optimisation and fabrication of SU-8 based electro-thermal micro-grippers. *J. Micro-Nano Mech.*, **6**, 3–22.

[12] Volland, B.E., Heerlein, H., and Rangelow, I.W. (2002) Electrostatically driven microgripper. *Microelectron. Eng.*, **61/62**, 1015–1023.

[13] Chen, T., Chen, L., Sun, L., and Li, X. (2009) Design and fabrication of a four-arm-structure MEMS gripper. *IEEE Trans. Ind. Electron.*, **56** (4), 996–1004.

[14] Bazaz, S.A., Khan, F., and Shakoor, R.I. (2011) Design, simulation and testing of eleltrostatic SOI MUMPs based microgripper integrated with capacitive contact sensor. *Sensor Actuat. A-Phys.*, **167** (1), 44–53.

[15] Chang, H., Zhao, H., Ye, F., Yuan, G., Xie, J., Kraft, M., *et al.* (2014) A rotary comb-actuated microgripper with a large displacement range. *Microsyst. Technol.*, **20** (1), 119–126.

[16] Beyeler, F., Neild, A., Oberti, S., Bell, D.J., Sun, Y., Dual, J., *et al.* (2007) Monolithically fabricated microgripper with integrated force sensor for manipulating microobjects and biological cells aligned in an ultrasonic field. *J. Microelectromech. Syst.*, **16** (1), 7–15.

[17] Piriyanont, B. and Moheimani, S.O.R. (2014) MEMS rotary microgripper with integrated electrothermal force sensor. *J. Microelectromech. Syst.*, **23** (6), 1249–1251.

[18] Estevez, P., Bank, J.M., Porta, M., Wei, J., Sarro, P.M., Tichem, M., *et al.* (2012) 6 DOF force and torque sensor for micro-manipulation applications. *Sens. Actuator A: Phys.*, **186**, 86–93.

[19] Wei, J., Porta, M., Tichem, M., and Sarro, P.M. (2009) A contact position detection and interaction force monitoring sensor for micro-assembly applications, in *Proc. 15th Int. Conf. on Solid State Sensors, Actuators and Microsystems (Transducers 2009)*, Denver, CO, pp. 2385–2388.

[20] Grade, J.D., Jerman, H., and Kenny, T.W. (2003) Design of large deflection electrostatic actuators. *J. Microelectromech. Syst.*, **12** (3), 335–343.

[21] Krijnen, B. and Brouwer, D.M. (2014) Flexures for large stroke electrostatic actuation in MEMS. *J. Micromech. Microeng.*, **24** (1), 015 006.

[22] Xu, Q. (2015) Design, fabrication and testing of a MEMS microgripper with dual-axis force sensor. *IEEE Sensors J.*, **15** (10), 6017–6026.

Index

Abbe errors, 80
Actuation force, 78, 100
Actuation stiffness, 98, 101, 248
Actuator decouplers, 47
Adhesive force, 262
Adhesive layers, 131
Amplification ratio, 98, 101, 248, 260
Analytical models, 52, 77, 146, 165, 185, 208, 223, 252
Area ratio, 45, 55, 68, 71, 88
Atomic force microscope (AFM), 10, 93
Axial load, 75
Axial-symmetric structure, 71

Bandwidth, 64, 158, 207
Bending, 49
Bending deflection, 142
Bending deformations, 27, 51, 73, 77, 115, 148, 174, 191, 225
Bidirectional drive, 214
Bidirectional motion, 131
Bidirectional translation, 115, 136
Blocking force, 100, 120
Bode plots, 129, 154, 180
Buckling, 49, 50, 72, 75
Buckling deformation, 75
Buckling effect, 51, 89
Butterworth filter, 203

CAD model, 79, 126, 201, 222, 252
Calibration, 127, 202, 228, 258

Capacitive force sensor, 245
Capacitive sensors, 12, 245
Capacitive-to-voltage converter, 249
Cartwheel flexure, 171
Center shift, 181
Centimeter range, 77, 81, 89, 100
Characteristic equation, 76
Characteristic radius, 27
Chattering, 38
Circular contouring, 84, 156
Clearance adjuster, 220
Closed-loop control, 55, 203
Coarse resolution, 132, 241
Coarse stage, 93, 94
Coarse/fine resolution ratio, 114, 117, 145, 146, 159
Comb drives, 249
Compactness, 4, 55, 97
Compactness requirement, 23, 164, 192
Compensation factor, 78
Compliance matrix, 187
Compliant gripper, 213, 214, 245
Compliant guiding bearing, 115
Compliant mechanisms, 1, 213
Compound parallelogram flexure (CPF), 4
Compound radial flexure (CRF), 9, 164
Compressive force, 136
Compressive loads, 49, 75
Computer vision, 214
Computer-aided design (CAD), 4
Conceptual design, 99, 115, 136

Design and Implementation of Large-Range Compliant Micropositioning Systems, First Edition. Qingsong Xu.
© 2016 John Wiley & Sons Singapore Pte Ltd. Published 2016 by John Wiley & Sons Singapore Pte Ltd.

Confidence intervals, 84
Confidence levels, 84
Consistency, 231
Contact detection, 235
Contact force, 214
Contouring accuracy, 86
Control bandwidth, 86, 132, 180
Control simulation, 61, 179
Control technique, 12
Convergence, 61
Cooperative positioning, 156
Coordinate transformations, 187
Corner-filleted hinge, 169
Cost function, 59
Coupling effect, 82, 83, 94
CPF, 21
CRFs, 186
Critical length, 49, 75
Critical load, 50, 75
Crosstalk, 54, 71, 81
Cutoff frequencies, 86
Cylindrical constraint, 124

Damping coefficient, 96
Damping effect, 203
Damping ratio, 203, 206
Data acquisition, 34, 126, 202
Deamplification, 3
Decoupled motion, 89
Decoupling design, 93, 111
Deep reactive ion etching (DRIE), 254
Degree-of-freedom (DOF), 54
Depth of view, 214
Design criteria, 146
Differential equations, 96
Digital microscope, 202
Digital PID, 57
Digital PID control, 39, 109
Discrete-time sliding mode control
    (DSMC), 13
Discrete-time system model, 64
Displacement amplification, 260
Displacement amplifier, 97, 107
Displacement sensors, 181
Disturbances, 13, 43
Drift, 262

Driving displacement, 248
Driving distance, 216
Driving force, 121, 136, 138, 171, 217, 247
DSMC, 39, 42
Dual-actuation stage, 100
Dual-range motion, 115
Dual-range stages, 114
Dual-resolution stage, 116
Dual-servo control, 109
Dual-servo stage, 11, 114
Dual-servo system (DSS), 10, 93
Dynamic characteristics, 237
Dynamic crosstalk, 82
Dynamics equation, 76
Dynamics model, 36, 95
Dynamics performance, 31, 35, 53, 81, 103,
    107, 124, 129, 132, 148, 154, 174,
    178, 199, 206, 225, 238, 254

Eigenvalue, 61, 76
Eigenvector, 76
Elastic buckling, 49, 75
Elastic deformation, 2, 223
Elastic energy, 100
Electric noise, 150
Electrical discharge machining (EDM), 12,
    33, 78, 104, 126, 149, 169
Electromagnetic actuator, 214, 245
Electrostatic actuator, 245
Electrostatic force, 249
Electrothermal actuator, 245
EMPC, 55, 58, 64, 66
Enhanced model predictive control
    (EMPC), 46
Environmental interaction, 241, 251
Equilibrium of moment, 168
Equivalent mass, 28
Equivalent stiffness, 115, 117, 136, 141,
    217, 247, 250
Estimation, 59

Fabrication errors, 158, 207
Fabrication imperfection, 258
Fast Fourier transform (FFT), 63
Fatigue analysis, 201, 226
Fatigue failure, 201

Fatigue lifecycle, 208
FEA, 30, 52, 73, 76
FEA simulations, 101, 124, 146, 147, 164, 173, 197, 208, 222
Feedback control, 70, 111, 164
FFT, 178, 206
Field-programmable gate array (FPGA), 80
Fine resolution, 132, 241
Fine stage, 97, 98
Finite-element analysis (FEA), 93, 114, 185
Fixed–fixed constraints, 75
Fixed–guided constraint, 8
Fixed–guided flexures, 118, 139, 218
Fixed–guided beams, 141
Fixing schemes, 25, 192
Flexible element, 1
Flexure bearings, 2, 121, 136
Flexure hinges, 2, 68
Flexure mechanism, 47, 95, 214
Folded leaf springs, 99
Force control, 264
Force sensor, 213, 214, 231
Force wrench, 187
Force–sensing gripper, 264
Free-body diagram, 168, 191
Free-motion dynamics, 29
Frequency response method, 35, 61, 81, 86, 107, 129, 154, 178, 206, 237
Frequency responses, 61, 63, 81, 107, 178, 180

Gauge factor, 121, 144, 195, 220
Gripping force, 245, 250
Gripping force sensing, 214
Gripping force sensor, 263
Gripping motion, 214, 216
Gripping range, 255
Guiding mechanism, 95

Half-bridge circuit, 195, 202
High bandwidth, 75
Histogram, 84, 110, 129, 151, 204, 229
Horizontal interaction, 236
Hybrid micropositioning, 1
Hysteresis, 34, 55, 106, 181, 205, 260

Hysteresis effect, 34, 81
Hysteretic nonlinearity, 34, 111

Impact effect, 132
In-plane motion, 31
In-plane rotation, 76
Incremental control, 59, 65
Inertia effect, 239
Input decoupling, 47, 71
Interaction, 97, 111, 245
Interaction effect, 114
Interaction force, 98, 214, 216, 235, 245, 246, 251, 261
Interaction force reducer, 98
Interaction force sensing, 214, 245
Interaction force sensor, 263
Interaction stiffness, 252
Interference, 11, 83, 94, 102, 158
Interference behavior, 94, 96, 103, 109, 114
Interference motion, 103
Inverted microscope, 257

Kinematic scheme, 4
Kinetic energy, 27

Laplace transform, 96
Large-range gripper, 215
Laser displacement sensor, 12, 34, 54, 150, 177, 228
Laser interferometer, 12, 42
Lateral comb drive, 245
Lateral displacement, 258
Leaf flexures, 2, 23, 75, 114, 117, 139, 165, 187, 216, 247
Leaf springs, 49, 68
Life cycles, 227
Linear actuator, 114
Linear guiding flexure, 216
Linear model, 64
Linear regression, 260
Linear springs, 220
Linear-time-invariant (LTI), 56
Lorentz force law, 12
Low-pass filters, 89, 158
Luenberger observer, 60

Magnitude responses, 86
Manufacturing tolerance, 9, 77, 131, 192,
    203, 206, 229, 238
Mass–spring–damper system, 36
Maximum allowable stress, 101
Maximum stress, 169, 173, 192, 195, 223
MCPF, 21, 23, 46, 71, 119, 139, 141
MCPFs, 104, 214, 216
MCRF, 165, 167, 191, 217
MCRFs, 186, 207, 214, 245
Measuring principle, 172
Mechanical stoppers, 88, 115, 193, 219
MEMS, 207
MEMS microgripper, 245, 264
Mesh density, 197
Mesh model, 197
Mesh size, 197
Micro-/nanomanipulation, 45
Microassembly, 213, 263
Microbalance, 236
Microelectromechanical systems (MEMS),
    3, 244
Microfabrication, 254
Microforce sensing probe, 258
Microgripper, 132, 244
Microhandling, 213
Micromanipulation, 213, 244
Micromanipulator, 238, 255, 261
Micropositioning, 1
Micropositioning stages, 2
Micropositioning systems, 1
MIMO control, 114
Minimum-stiffness requirement, 23, 164,
    169, 192
Mirror-symmetric stage, 47
Mirror-symmetric structure, 71
Modal analysis, 31, 53, 124, 148, 174, 199,
    225, 254
Mode frequencies, 31
Mode shapes, 31, 32, 54, 149, 174, 200
Model predictive control (MPC), 13
Model uncertainties, 13
Model-free control, 13
Model-free controller, 39, 57, 109
Modular design, 71
Moment of inertia, 75, 99, 167, 190

Monolithic stage, 70, 71
Monolithic structure, 8, 70
Motion control, 63
Motion decoupling, 55, 70, 101
Motion guiding mechanism, 2
Motion range, 73, 78, 104, 116, 120, 131,
    138, 181
Motion tracking, 180
Motion twist, 187
MPC, 46, 58
Multi-resolution positioning, 158
Multi-resolution stage, 138
Multi-stage compound parallelogram
    flexure (MCPF), 4
Multi-stage compound radial flexure
    (MCRF), 9, 164, 186
Multi-stroke micropositioning, 10, 136
Multi-stroke stages, 135
Multiple-input/multiple-output (MIMO), 94

Nanopositioner, 231, 258
Nanopositioning stage, 10, 11, 93
Natural frequency, 27, 29, 75, 201
Negative stiffness, 249
NMP, 57
NMP plant, 63
NMP system, 46
Noise, 41, 151, 203, 228, 260
Nominal parameters, 77
Nominal system, 57
Non-minimum-phase (NMP), 46
Non-minimum-phase systems, 68
Normal distribution, 129, 151, 228

Observer, 109
Open-loop control, 55, 81
Open-loop performance, 207
Optical encoders, 207
Optimization, 29, 59, 264
Optimum design, 55
Oscillation, 64, 236
Out-of-plane displacement, 76
Out-of-plane motion, 175
Out-of-plane payload, 76, 89, 174, 175,
    193, 199
Out-of-plane stiffness, 175

Output decoupling, 6, 47
Overconstraint, 164
Overlay image, 203
Overshoot, 66, 84, 110, 180, 203

Parallel connections, 142
Parallel mechanism, 46, 47, 71
Parallel plates, 158, 247
Parallel-kinematic architecture, 88
Parallel-kinematic scheme, 5, 70
Parallelogram flexure, 3, 95
Parametric design, 73, 146
Parasitic motion, 26, 47, 71, 73, 80, 81, 187
Parasitic translation, 3, 165
Payload capability, 175, 207
Perturbation estimation, 36
Perturbations, 57
Pessen integral rule, 41
Phase portrait, 39
Physical model, 95
Pick-and-place operation, 213
PID, 37, 42, 164
PID control, 39, 57, 66, 83, 89
PID control gains, 110, 180
PID controller, 156, 178
Piezoelectric stack actuators (PSAs), 11, 164
Piezoresistive, 228, 232
Piezoresistive sensor, 138, 158, 214
Plastic deformation, 119
Pole placement, 64
Positioning accuracy, 181
Positioning error, 111
Positioning performance, 89
Positioning resolution, 204, 207
Potential energy, 27
Precision positioning, 6, 38, 71, 81, 113, 132
Prediction horizon, 58, 65
Preloading effect, 107
Preview, 67
Printed circuit board (PCB), 254
Prismatic hinge, 7
Proof-of-concept design, 133, 159
Proportional-integral-derivative (PID) control, 13

PSA, 88, 93, 106
Pseudo-rigid-body (PRB) model, 27
Pull-in stability, 264

Quantitative models, 30, 52
Quarter-bridge circuit, 149, 220, 242
Quarter-bridge circuits, 121, 158

Radial flexure, 8, 164, 186
Rapid prototyping (RP), 12
Reachable workspace, 73, 88, 89
Real-time control, 54, 80, 127, 177, 228
Real-time controller, 202
Reliability, 39
Resistive sensor, 135, 138
Resolution, 41, 150
Resonant frequencies, 35, 76, 82, 107, 124, 154, 175, 200, 254
Resonant modes, 31, 103, 124, 148, 254
Resonant peak, 206, 237
Revolute hinge, 7
Revolute joint, 2
Right-circular hinges, 95
RMSE, 41, 110, 180
Robotic gripper, 213
Robust control, 13, 46
Root-mean-square (RMS), 228
Root-mean-square error (RMSE), 39, 64
Rotary angle, 8
Rotary bearing, 216
Rotary guiding, 214
Rotary motion, 8
Rotary motor, 185
Rotary positioning, 8
Rotary springs, 207
Rotary stages, 8, 163
Rotary VCM, 194, 201
Rotation center, 184
Rotational angle, 185, 190
Rotational bearings, 8
Rotational micropositioning, 1
Rotational motion, 124
Rotational range, 164, 171
Rotational stiffness, 31

Safety factor, 88, 101, 197
Sampling frequency, 61, 64
Sampling rate, 34, 202
Sampling time, 57, 64, 110
Saturation functions, 110
Scanning electron microscope (SEM), 254
Scanning probe microscopy (SPM), 10
Screw theory, 187
Sensing scheme, 172
Sensitivity, 185, 219, 228, 261
Sensitivity ratio, 229
Sensor sensitivity, 196, 203
Serial-kinematic scheme, 4
Set-point positioning, 39, 64, 84, 110, 180
Settling time, 110, 180
Shear modulus, 187
Side instability, 249
Side sticking, 258
Signal conditioning, 232
Signal processing, 207
Signal-to-noise ratio (SNR), 117, 123, 138, 146
Silicon-on-insulator (SOI), 244
Simulation, 66, 101
Single-drive stage, 114
Single-input/single-output (SISO), 56, 71
Sinusoidal signal, 129, 152, 228
Sinusoidal trajectory, 180
SISO, 93
SISO control, 94, 113
SISO controller, 63, 83
Skew-symmetric matrix, 188
Sliding function, 37
Sliding mode control (SMC), 13
SNR, 121, 128, 145
Soft mechanisms, 12
SOI, 245
SOI wafer, 254
Spectral analysis, 81
Spherical joint, 2
Spring constant, 168, 191, 217, 247
Stability analysis, 37
Standard deviation, 129, 151, 204
State observer, 13, 60, 64
State-space equation, 58
State-space model, 64

Static crosstalk, 82
Static structural analysis, 74, 76, 147, 173, 197
Statics performance, 101, 106, 124, 147, 150, 173, 197, 223, 252
Steady-state error, 46, 64, 66, 110
Step response, 66
Stick–slip actuators, 11
Stiffness, 24, 36, 74, 97, 98, 241
Stiffness matrix, 187
Stiffness model, 51
Strain gauges, 121, 128, 144, 149, 222
Strain sensors, 151
Strain-gauge sensors, 12, 114, 128, 144, 194
Stress concentration, 2, 169
Stress life analysis, 201
Stress stiffening, 7, 164
Stroke stopper, 136, 143
Sub-micrometer, 21, 70
Sub-micrometer accuracy, 89
Sub-micrometer resolution, 42
Swept-sine signal, 129, 154, 178, 206
Swept-sine waves, 35, 61, 81, 237
Symbolic formulation, 187
Symbolic model, 198
System identification, 57

Tangential direction, 185, 216
Tangential translation, 167, 190
Thermal expansion, 25
Topology optimization, 2
Torsional stiffness, 184, 189, 194, 198, 217, 247
Total decoupling, 6, 47, 70, 71
Transfer function, 63
Transformation matrix, 188
Transient response, 46, 66, 94, 180
Transient response speed, 111
Translational displacement, 190
Translational micropositioning, 1, 45
Transverse comb drive, 245
Transverse displacement, 71, 99, 258
Transverse load, 47, 71
Transverse motion, 103
Transverse stiffness, 25, 47, 71, 98, 103, 104, 117, 235, 249

Two-dimensional translation, 71
Two-layer structure, 88
Two-stage force sensing, 215

Uncertainties, 43
Uniaxial flexure stage, 25
Universal hinge, 7, 207
Universal joint, 2

Variable-stiffness mechanism, 114, 135
VCM, 33, 51, 54, 80, 93, 106, 126, 149,
     170, 177, 227
Vertical interaction, 237
Virtual work principle, 97

Visual servoing, 207
Voice coil motor (VCM), 11, 145, 164, 185

Wheatstone bridge, 126, 144
Wire-EDM, 80, 176, 201, 227

Yield strength, 2, 23, 101, 142, 168, 197,
     218
Young's modulus, 23, 98, 119, 141, 167,
     187, 195, 217, 247

Z–N method, 41, 64, 84, 180
Ziegler–Nichols (Z–N), 39, 58, 110